# Crossing the Neoliberal Line

*Pacific Rim Migration and the Metropolis*

In the series

PLACE, CULTURE, AND POLITICS
edited by Neil Smith

ALSO IN THE SERIES:
George L. Henderson, *California and the Fictions of Capital*

# Crossing the Neoliberal Line

*Pacific Rim Migration and the Metropolis*

KATHARYNE MITCHELL

TEMPLE UNIVERSITY PRESS
Philadelphia

Temple University Press
1601 North Broad Street
Philadelphia PA 19122
www.temple.edu/tempress

⊗ The paper used in this publication meets the requirements of the American
National Standard for Information Sciences—Permanence of Paper for Printed
Library Materials, ANSI Z39.48-1992

Library of Congress Cataloging-in-Publication Data

Mitchell, Katharyne.
  Crossing the neoliberal line : Pacific rim migration and the metropolis /
Katharyne Mitchell
    p.  cm. — (Place, culture, and politics)
  Includes bibliographical references and index.
  ISBN 1-59213-083-6 (cloth : alk. paper) — ISBN 1-59213-084-4 (pbk. :
alk. paper)
    1. Civil society—Case studies.  2. Liberalism.  3. Emigration and immi-
gration—Social aspects—Case studies.  4. Chinese—British Columbia—
Vancouver.  5. Hong Kong (China)—Emigration and immigration—Social
aspects.  6. Vancouver (B.C.)—Emigration and immigration—Social aspects.
7. Civil society—British Columbia—Vancouver.  8. Liberalism—British
Columbia—Vancouver.  I. Title.  II. Series.

JC337.M58 2004
304.8'711205125–dc22                                              2004044018

2  4  6  8  9  7  5  3  1

*To Matt*

# Contents

# Acknowledgments

THIS BOOK has been constructed and reconstructed over the course of nearly a decade, and hence the debts have piled up in a somewhat embarrassing collection. The research presented here was conducted primarily in 1990 and 1991, although I have made repeated forays back into Vancouver life, at least partly because it is such an irresistibly beautiful city. Early research was funded by a Fulbright Institute of International Education grant, the University of California at Berkeley, and the Pacific Cultural Foundation of Taiwan. I have also received assistance from the Royalty Research Fund at the University of Washington.

Intellectual debts are more difficult to pin down, since this work draws on the musings of friends, students, mentors, and colleagues in various geographical sites over the course of an academic lifetime. Allan Pred's ongoing mentorship and friendship remain an enduring source of support, long after my Berkeley years, and I am happy to finally have a public venue in which to thank him. I have also enjoyed stimulating conversations and not a few beers with Michael Watts, Dick Walker, and Gill Hart. At the University of British Columbia, I received much-needed help from David Ley, who has always taken the time to talk with me about Vancouver politics and urban transformation. I am grateful also to Derek Gregory, Gerry Pratt, Cole Harris, Trevor Barnes, and Dan Hiebert. Enduring friendships made in Berkeley and Vancouver have fostered many of the ideas presented here. Thanks, in particular, go to Kris Olds, Bruce Braun, Jennifer Hyndman, Rod Neumann, Rick Schroeder, Susan Craddock, George Henderson, and Krystyna von Henneberg.

In Vancouver and Hong Kong many people offered assistance on this project, most of whom I quote anonymously in this text, but all of whom I remember for their enthusiasm and helpfulness. I would like to thank, in particular, Annie Leung and Betty Tangye, who spent tens of hours talking with me, who directed me to other people to interview, and who shared boxes and boxes of newspaper clippings and other collected materials. They, and many others, made this research not just possible, but tremendously enjoyable.

Over the past decade my life has been immeasurably brightened by my wonderful colleagues Vicky Lawson and Lucy Jarosz. They have made the third floor of Smith Hall a warm, vibrant, and fun space, and I am endlessly grateful that I have such kind and smart women working just next door. In recent years I have also enjoyed stimulating conversations with newer members of the department—Suzanne Davies Withers, Mark Ellis, Steve Herbert, and Kim England. Thanks also go to Marv and Penny Waterstone for their wonderful hospitality at the Waldport think tank, where we go annually to get tanked on some very nice red wine.

I have been greatly sustained over the last several years by the tremendous intellectual energy and dynamism of Cindi Katz and Sallie Marston. Working with them on our coedited book, *Life's Work,* made me think about transnationalism and questions of social reproduction in new ways and enriched many of the ideas presented here. Even more important, their friendship seems to transform even mundane tasks and dull conferences into something sparkly.

I am grateful to those who took precious time off from their own work to read parts of this manuscript and give much needed advice. Many thanks to David Ley, Sallie Marston, Michael Peter Smith, Vicky Lawson, Lucy Jarosz, Matt Sparke, Linda Nash, Mark Ellis, and Sarah Stein. Special thanks in this regard go to Peter Wissoker and most especially Neil Smith for their major editorial assistance on the entire manuscript.

My parents have been staunch supporters of all my scholarly endeavors. My heartfelt thanks to Beth and George Mitchell for their love and assistance over the years. Special thanks also go to Gary and Eleanor Hamilton and to Becca Sheuerman, perennial bright spots in my emotional and intellectual landscape.

The project's lengthy undertaking can be traced almost directly to my children, Sage and Emma, who have never manifested a shred of patience, and who have not once allowed me to put them on hold while I finished just one more paragraph. I want to thank them, in particular, for giving me a slower book but a richer life.

Matt Sparke has been my true companion and comrade for quite a few years now. My work has benefited from his keen critiques and my soul has benefited from his love and affection. Our Friday nights of champagne and oysters together have sustained me through even the darkest and wettest of Seattle winters. Here's a toast to many more.

Parts of the text have been adapted from four of my earlier publications:

"Visions of Vancouver: Ideology, Democracy, and the Future of Urban Development." Reprinted with permission from *Urban Geography*, Vol. 17, No. 6 (1996), pp. 478–501. © V. H. Winston & Son, Inc., 360 South Ocean Boulevard, Palm Beach, FL 33480. All rights reserved.

"Conflicting Geographies of Democracy and the Public Sphere in Vancouver, B.C." *Transactions of the Institute of British Geographers*, Vol. 22, No. 2 (1997), pp. 162–179. Reprinted with permission of Blackwell Publishing Ltd.

"Transnationalism, Neo-liberalism and the Rise of the Shadow State." *Economy and Society*, Vol. 30, No. 1 (2001), pp. 165–189. Reprinted courtesy of Taylor & Francis, http://www.tandf.co.uk.

"Multiculturalism, or the United Colors of Capitalism?" *Antipode*, Vol. 25, No. 4 (1993), pp. 263–294.

# Crossing the Neoliberal Line
*Pacific Rim Migration and the Metropolis*

# 1    Introduction

## *Neo/Liberal Disjunctures*

FROM THE VANTAGE POINT of the stars in 1989, the transformation toward a newly forming "global liberalism" on planet earth seemed ubiquitous and triumphal. From the fall of the Berlin Wall, to regime changes in Poland and Romania, to glasnost, to the end of the Cold War, neoliberalism and the market appeared inevitable and unstoppable, a smug and self-sufficient duo with no barriers in sight, global coverage the ultimate goal, and victory the annihilation of all alternate forms of social and economic organization. An unflagging belief in the economic logic of laissez-faire capitalism seemed to be the new global hegemony, and it was a logic that appeared perfect, infallible. Coronil captured this perspective well:

> As an expression of this millennial fantasy, corporate discourses of globalization evoke with particular force the advent of a new epoch free from the limitations of the past. Their image of globalization offers the promise of a unified humanity no longer divided by East and West, North and South, Europe and its Others, the rich and the poor. As if they were underwritten by the desire to erase the scars of a conflictual past or to bring it to a harmonious end, these discourses set in motion the belief that the separate histories, geographies, and cultures that have divided humanity are now being brought together by the warm embrace of globalization, understood as a progressive process of planetary integration.[1]

It is this global village image of harmonious integration that was harnessed and put to work for corporations from AT&T to Benetton, from Global Crossing to IBM. Different histories and bodies were fused into a perfect oneness, and then difference itself was banished from future discussion. This remains the ultimate fantasy of disembodied, ahistorical spacelessness, where there are no material or discursive frictions to hinder the space of flows. It is the fantasy of neoliberalism, a world without historical and spatial

1

closure, without fissures and alienation; it is the transcendent moment for capitalism as well, for in a world without "the scars of a conflictual past" or "the separate histories, geographies, and cultures that have divided humanity," there also exist no barriers to the free circulation of capital and laboring bodies across space.

For the corporate world and for neoliberal thinkers like Francis Fukuyama, human society is poised for a great evolutionary leap—the leap into total obeisance to the laws of capitalist motion. For Fukuyama, the "end of history" means the end of the twentieth-century experiments of different doctrines of government; it means the victory of capitalism and modern, liberal freedoms over socialism, communism, fascism, and dictatorship.[2] For him, the current apocalypse of "the end" is also the current utopia—the moment where the limits of so-called social engineering are made manifest, and a *neo*liberal, market-based order grounded in human nature emerges triumphant. Fukuyama's "new" liberalism exists in profound contrast to a social, democratic, and interventionist liberalism of the mid-twentieth-century era. It is a philosophy of noninterference and of the triumphant individual, incorporating notions of freedom and autonomy across scales, from the individual human body to global corporations. The imagined neoliberal world is one of purity and flow, a world of perfect individuals and markets engaging and contracting with each other, unobstructed by the depredations of the state. The end of history for Fukuyama and his numerous neoliberal compadres means the final moment of evolution in the Hegelian sense of a progressive development of political and economic institutions. In this contemporary picture, the "liberal" national state of the Keynesian moment will be transcended and surpassed by the "neoliberal" postnational state of the global era. Economic globalization, finally made possible through the perseverance of laissez-faire capitalism's dynamic and inexorable motion, will mean the end of old, static, and socially proscribed traditions and practices, and the beginning of a new world order.

But wait. If for a moment we remove ourselves to a less lofty vantage point, that of a small city on one edge of the former British Empire, this assumed neoliberal future becomes a little more dubious. In 1989 this city was transforming almost before our eyes from a provincial backwater, an achingly beautiful yet placid city, into a global metropolis, a gateway between West and East. In this year, just one year after the Hong Kong property magnate Li Ka-Shing purchased the former Expo '86 lands for a negligible sum, property in Vancouver, British Columbia, was the hottest real estate in the world.[3] It was a city on fire.

In 1989, Vancouver was rapidly and unceremoniously swept into a "progressive process of planetary integration" or, if not that, certainly into a new form of global currency. But there was trouble brewing in paradise—over changes in house styles and house prices and who was moving in next door, over landscapes and tree removal and the cost of renting an apartment, over neighborhood character and zoning amendments, and over who was making the decisions about these things and why. Minor troubles perhaps in the greater panworld vision of progressive planetary embrace, but in this book I show how these erstwhile local and confined tactics and struggles over the production of space matter, and matter in some large ways.

The main argument in this book runs along the following lines: certain kinds of global flows—for example, the flows of wealthy transnational migrants and their capital—are central to neoliberal state formations but also are deeply disruptive to national liberal, social, and political narratives as they have developed and become embedded through time in the crusty layers of urban social life. As a result, deep and abiding tensions have developed between state practices that facilitate these flows and render them abstract in rhetorical and spatial terms, and national norms that relate to the localized, territorial, and embodied practices and understandings of urban civil society. In Canada these tensions manifest some of the problems neoliberal politicians and planners have encountered when trying to roll out a new vision of warm planetary embrace, only to be caught in the sticky geographical banalities of everyday life.

Between Hong Kong and Vancouver in the final two decades of the twentieth century, there was a movement of people and capital unprecedented in scope.[4] This movement was part of a vast emigration of middle-class and wealthy Chinese residents out of Hong Kong and into a number of cities worldwide, a direct consequence of the pending transfer of British colonial control to China in 1997. Two key events spurred the outflow of both people and capital: first, the signing of the Sino-British Joint Declaration by Prime Minister Margaret Thatcher in 1984, which indicated to the world that the transition to mainland-Chinese control would actually occur; and second, the massacre of students and workers at Tienanmen Square in 1989, which indicated that this transition would not take place without considerable human and financial costs. One key event directed many of the wealthiest among this group to Canada. This was the initiation of a new immigration law entitling those with capital or business experience to skip processing queues, procure landed-immigrant status, and "join" the greater Canadian community without the usual hurdles and aggravating bureaucratic delays.[5]

In the process of moving into and working through the urban spaces of Canadian society, the Hong Kong immigrant entrepreneurs challenged the implicit assumptions of British liberalism as they were adapted and transformed in Canada. Because of the vast wealth and cosmopolitan savvy of many of the immigrants who landed in Vancouver, they were able to contest, both discursively and by their very nonwhite presence in white neighborhoods, many of the normative assumptions of what constituted Canadian liberalism and national identity.[6] They also challenged, implicitly and explicitly, national narratives of tolerance, rationality, universality, normality, and harmony associated with legitimacy and consent in governance, narratives foundational to the premise of a liberal Western nation.

These challenges were conspicuous and had some purchase in society primarily because of the privileged economic position of the immigrants, which made them simultaneously symbolic "carriers" of capital and also much more visible and powerful than poor migrants. In a lengthy exegesis on the problem of ideology in Marx's thought, Stuart Hall discussed the invisibility of certain key moments of the circuit of capital owing to the fetishization of the market and the lack of linguistic concepts and categories that might lend insight into the production process as a whole.[7] I argue that, in complex ways, the heretofore "invisible" workings of certain aspects of the circuits of capital in Vancouver were rendered visible (vis-à-vis the practical consciousness of the residents) as a result of the spatial practices and presence of the Hong Kong immigrants, who were positioned as the ultimate vectors of fast capital. This sudden visibility forced Canadian residents to consider more fully the repercussions of the state's neoliberal agenda, which facilitated the creation of a spaceless, free-flowing, giant enterprise zone, especially with the booming Pacific Rim economies. At the same time, it exposed the limits of many taken-for-granted liberal assumptions in Canadian society, as well as the rapid hemorrhaging of liberalism's twentieth-century welfarist protections, as residents, politicians and immigrants simultaneously clashed over the meanings and implementation of liberalism *and* neoliberalism in space.

Documenting in detail the transformation of space and consciousness in a particular urban environment makes it possible to understand the tightly interwoven relationship between socioeconomic change, urban spatial transformation, and the narratives and practices of contemporary regimes of governance. As developers, long-term residents, politicians, and the Hong Kong Chinese immigrants struggled over the local and national spaces of Canadian society, liberalism was literally contested in the streets. What Uday Mehta calls the "synaptic links in liberal thinking" were thereby exposed, straining conscious thought "to the point where liberals [were] forced to bring to mind what otherwise remained unconscious."[8] The sotto voce language of reason, tolerance, and normality could no longer obscure the inherent exclusions and limitations within the narratives of multiculturalism, urban land governance, democracy, domesticity, and other touchstones of Western liberalism. The "wrong" bodies in the "wrong" places forced the whole panoply of ideological apparatuses out into the open. They rendered visible and public, for a time, what had formerly remained implicit, private, and assumed. The global and transnational nature of these particular "unfamiliar" bodies was crucial in this disruptive moment.

However, the economically liberal rhetoric of private property and individual rights expressed by many of the upper-class immigrants from Hong Kong provided a critique of contemporary Canadian civil society from a position that was by no means disruptive to the state's deepening neoliberal agenda. Thus, while many liberal social conventions of the twentieth century were shown to mask ongoing racial inequities stemming from their historical provenance in British liberal thought, the *class-based* inequities of early economic liberalism were generally strengthened with the arrival of these immigrant-entrepreneurs. Contemporary social liberalism in Canada became vulnerable to the challenges posed by the wealthy and middle-class immigrants for two reasons: first, the constitution of the early British liberalism of Locke, the Mills, Macaulay, Maine, and Bentham as a fundamentally exclusionary and aggressive doctrine (despite its stance of inclusive neutrality);[9] second, the increasingly visible sedimentation of those exclusions and aggressions in space. The immigrants' and long-term residents' struggles over space highlighted the tension between an economically based global agenda of *economic* liberalism, harkening back to the earliest conceptualizations of the term, and a more recent national narrative of *social* liberalism, deriving from the context of twentieth-century welfarism and the claims of contemporary citizenship.

As is probably already evident, this book engages with a number of disparate literatures. My intention is not to intervene abstractly in philosophical debates, but to juxtapose my empirical work with those debates and bring them into tension with each other. If you ask me what is the object of my work, like Hall, "the object of the work is to always reproduce the concrete in thought—not to generate another good theory, but to give a better theorized account of concrete historical reality. This is not an antitheoretical stance. I need theory in order to do this. But the goal is to understand the situation you started out with better than before."[10] In addition, the theoretical frames that interest me most—those relating to liberalism and neoliberalism, transnationalism and globalization, and hegemony—are not literatures customarily put into conversation with each other, and certainly do not often engage with space in more than a metaphorical sense. In this work I ask how new kinds of transnational movements are putting pressure on liberal aspirations and self-conceptions—in particular, how the "spacelessness" of classical liberal thought (one of the cornerstones of *neo*liberalism) is challenged through the production of space by "outsiders" who play by a slightly different and unfamiliar set of rules. I also want to juxtapose the question of hegemony with the facts of globalization, to ask what happens to our conceptualization of hegemony when it is no longer contained by the traditional nation format.[11] Finally, I take the question of space to heart. I consider through an emphasis on the interrelations of spatial scales how space (or its crucial absence) is deeply insinuated in the practices and policies of all varieties of liberal thought.

Aihwa Ong has engaged with similar themes in her thought-provoking book *Flexible Citizenship*, which documented the movements of wealthy Chinese transnationals and the shifting structures of power/knowledge in late capitalism.[12] While insightful and erudite with respect to questions of contemporary globalization, governmentality, deterritorialization, and the formation of alternative modernities, her narrative gives somewhat less attention to the processes of subjectivity formation for the transnational migrants themselves. Few locatable subjects are given much ethnographic substance or differentiation based on gender, class fraction, or urban and regional context. This lack leads to a different set of emphases in Ong's work.

For example, Ong's discussion of liberalism reflects this dearth of differentiated actors who form (or do not form) various kinds of social movements, who are divided (or not divided) by various kinds of class fractions or national origins or linguistic differences, and who resist (or do not resist) macro structures of authority. She writes of liberalism not as a political philosophy but as "an art of government" that includes an "array of rationalities whereby a liberal government attempts to resolve problems of how to govern society as a whole."[13] Through this privileging of the top-down, *state control* of philosophies such as liberalism, Ong becomes less attentive to the wide varieties of actually existing liberalism; the paradoxical tendencies between and struggles over economic, political, and social liberalism as they have developed through time and become encrusted in space and memory; and the different rhetorics and narratives of liberalism employed by the "subjects" of liberalism—people who live and work in cities that matter to them.[14] As with much of the literature on governmentality, the work produces an overly generalized and seamless account of a historical epoch, one that downplays the ambiguity and contingency of liberal and neoliberal projects and elides the messy actualities of resistance and rule.

As I show in this work, the spatial struggles over urban formation in Vancouver are directly implicated in the larger philosophical questions of regime change and of whose liberal "logic" becomes dominant. Clashes over which variety of liberalism takes precedence are rarely, if ever, decided by state decree, or by the outcome of regulatory strategies designed by the state to persuade its populations of the superiority of a particular kind of logic. The interesting ethnographic question to pursue, from my perspective, is not how subjects *adapt to* the contemporary problematics of time-space compression and transnational modalities. Rather it is how the *actions of* individual agents who negotiate the contradictory structures of late modernity consolidate and contest different logics of liberalism in space.

In this book I also engage with the question of transnationalism and its impact on national narratives and spaces. The spatial stretching across borders of transnational life is often theorized as one form of a national deterritorialization of both space and memory, leaving behind it a liberal policy vacuum that is then easily filled with the global-babble of neoliberalism. I argue instead that "respatialization" may be a more accurate term than "deterritorialization": it indicates a back-and-forth reworking of spatial arrangements and associated hegemonies, rather than a single, seemingly autonomous event in time and space (a space that is generally metaphorically framed, moreover). Framing the disjunctures of late modernity in terms of deterritorialization, as does the work of Appadurai, neglects the ways that processes of reterritorialization nearly always occur in conjunction with deterritorialization, and often serves to further entrench the project of global neoliberalism.[15] But more than this, it also positions hegemony (and counter-hegemonies) at the scale of the nation, rather than embracing a more Gramscian understanding of the multiple scales and spheres in which hegemonies are nourished and maintained. Although Appadurai intends to merely call attention to the spaces of potential intervention in hegemonic national narratives, the generally celebratory tone he employs throughout *Modernity at Large* indicates the desire to herald these deterritorializing disjunctures as the premier sites of resistance in a postnational era. Further, as Sparke points out, despite the use of the suffix "-scape" to describe five key disjunctures of the contemporary moment, the actual "land-scape" that serves as their referent is notably absent—aside from the usual ubiquitous but anemic spatial metaphors.[16]

Abstracting away from the specific contexts in which deterritorialization occurs limits the theorist's ability to chart shifting practices of deterritorialization and *re*territorialization under the advance of neoliberalism. Abstract analyses also tend to privilege macro formations that relate to state or economic articulations and dearticulations rather than examining the particular sociocultural configurations of power/knowledge that constitute and maintain them at a smaller scale. For example, as numerous scholars have discussed in recent years, an awareness of the social construction of scale gives insight into the manifold ways that different levels of governance are fluid and interlinked, and can morph into each other or "jump" from one to the other under certain conditions. Most research on scale emphasizes the role of the territorial state in the production of scalar fixes for capital, showing how and why the dominant scale of governance switches from national to local or supranational scales in periods of crisis.[17] Recent work by Marston, however, supplements this production-centered research by examining the ways in which social reproduction and consumption are also key processes in the construction of scale.[18]

Instead of focusing on the manner in which the state regulates and reconfigures scale for the purposes of capital accumulation, a focus that tends to position the state as unilinear and somewhat homogeneous, Marston emphasizes the multiple forces that constitute the state and affect its tactics and targets. One of the crucial ongoing factors for any given territorial state is legitimacy, which comes not only from the economic realm, such as effective capitalist management, but also from the cultural sphere, wherein the narratives of the nation predominate. Using a historical example from the progressive era in the United States, Marston demonstrates how women's struggles for collective consumption entitlements in the city eventually led to a shift in both the material practices and the ideological discourse of social service provisioning by the state. Employing a discourse of maternalism and

domesticity, the women successfully challenged the prior legit-
imizing narratives of capital and altered, in significant ways, the
practices of the state vis-à-vis the politics of scale.

This framing of scale construction with an eye toward the
hegemonic, as played out in particular context-specific struggles
over space, extends the question of state deterritorialization from
a purely economic one to a more complicated examination of the
ways that the state must negotiate the cultural tensions inherent in any liberal, democratic, and plural society. At the same
time, it is a reminder that hegemonic formations can be actively
unraveled by the collective agency of groups who manage, for a
time, to articulate their desires in a configuration of power that
works—that is to say, that effectively challenges normative
understandings of state-society relations. That these interventions are in no way automatically progressive merely because
they intervene in dominant narratives of nation or gender or race
should be understood from the start. Indeed, as Marston made
clear, the women who successfully fought for their rights to the
city did so on the backs of immigrants and other minorities.

In this book I regard the literal spatial positioning, movements,
and struggles of actors as integral to the conceptual analysis: first,
of how neoliberalism is both entrenched and resisted; and second,
of how modern liberal notions of reason, progress, equality, and
tolerance become used as rhetorical tactics in this larger hegemonic battle. The question of race is an abiding concern of my
analysis, reflecting its prominence in debates about the exclusions
inherent within liberalism, as well as the actual physical spaces
of inclusion and exclusion in cities and nations. Race is also shown
here as a discursive construction employed strategically by different actors, including those who have been, and continue to be,
racialized by dominant groups. The "self-orientalizing" moves by
some Chinese immigrants were part of these discursive tactics;[19]
they helped to dislodge cultural representations that had become
naturalized and fixed spatially over time.[20] As mentioned earlier,
however, although these reframing strategies may have actively
worked to reorganize the landscape and thus intervened in dominant representations of race, place, and nation, they were not *necessarily* progressive agendas, and in fact often helped to entrench
regressive socioeconomic formations such as neoliberalism.[21]

This work, then, is a spatial ethnography in the strongest sense, where space and the sedimented histories of life are not just theoretical variables but actively constitute what happens, how, and why.[22] In the following sections I touch on some of the theoretical literatures pertinent to this ethnography.

## TRANSNATIONALISM AND GLOBALIZATION

Transnationalization is one feature of the globalizing trends of the last few decades. These trends embrace changes in the systems of world governance, including the proliferation of regional and cross-border trade agreements and pacts, the end of the Cold War and the beginning of a so-called "new world order." They are also evident in economic shifts that involve the nature of capitalism as a global socioeconomic system, especially new geographies of production, trade, and consumption, the accelerating flows of capital and labor, the rise of new kinds of commodity flows, and the increasing polarization of wealth on both macro (geographic) and micro (household) scales.[23]

All of these processes were clearly evident to some degree in different historical epochs of the past, especially the end of the nineteenth century (although Abu-Lughod dates many of these processes as far back as the thirteenth century).[24] But what most globalization theorists argue is that in terms of the scale of transnational networking, the extension of global ties across space, the density of international relationships and partnerships, the economic concentration of wealth and power, and the general cultural awareness and ecumenality of society, contemporary globalization does indeed herald something new.[25] For people living through this historical moment, the *experience* of everyday life is qualitatively different.[26]

One of the reasons for the deeply visceral "experience" of globalization for many lies not just in the actual material changes of society but also in the inescapable rhetoric that has accompanied them. The narrative of globalization is so ubiquitous and relentless that it has taken on a life of its own and seems to be an actor that does things to people. Globalization thus is seen to cause people to lose their jobs and to become displaced from their homes. It is an alienated force that creates unemployment and dispossession. Because it is personless, however, this force can never be directly confronted. The narrative of globalization as agent thus releases corporations from responsibility and serves as a useful strategy for them in countering local resistance to unpopular managerial decisions. Beck writes of this common tactic: "All around, one hears the assertion that it is not corporate interests but 'globalization' which forces this or that painful break with the past. One of the 'laws' of the global market is that not-A must be done in order to achieve A: for example, that jobs must be axed or relocated in order to keep jobs safe where they are."[27]

Thus, although there are clearly a growing set of interdependencies between nations and regions, there is also a vast ideological apparatus associated with promoting the *idea* of interdependence. Further, the crucial aspect of this rhetoric of interdependence is not political (hence the current U.S. dismissiveness toward the United Nations and indeed any form of multilateralism), but *economic*. The rhetoric of globalism is the rhetoric of world market domination and the ostracizing, if not extermination, of all forms of government based on any alternative to this goal. Political action in other directions, especially in the realm of social reproduction, is attacked as diminishing the effectiveness, logic, and rationality of the laissez-faire marketplace.[28]

In this book I am interested in both the actual, material implications of transnational flows and networking, but also in the hegemonic production of and resistance to an ideology of world market domination, otherwise known as neoliberalism. It is the combination of the discursive framing of neoliberal globalization as necessary, inevitable, and beneficial, alongside the growth of global interdependencies in areas ranging from production, trade, finance and migration to cosmopolitan consciousness, ethnoscapes, the media, and culture, that makes contemporary globalization unique to this historical moment. Scholars of migration patterns, production complexes, and financial markets have pointed to the intricate, well-established linkages established between nations for centuries and claimed that globalization is just an old game with a few new tricks up its sleeve.[29] But even if we discount the differences in scale and volume of these historical connections, the overdetermined complex of multiple threads coming together, including the discursive, constitutes a new formation with a new logic that is qualitatively, as well as quantitatively, different from the articulations of the past.[30]

Holding in mind this material-discursive conjunction helps illuminate the ways in which the inexorable rhetoric of "globalization" has helped entrench the form of governance now known as neoliberalism. It also points to the sites of slippage when the rhetoric cannot hold, and where resistance gains ground.[31] In this work, I examine the literal sites in which the rhetoric of an inevitable and invincible neoliberal form of globalization is used to promote specific kinds of policies related to the production and dispensation of urban land. I also look at the actual places and moments when these discursive constructions are contested or fail, as counter-hegemonies are employed that call on notions of the liberal nation and the rooted, stable, and compassionate city.

One of the most important threads in the globalization tapestry of my narrative is the globalization of production and the attendant spatial shifts of both people and organizational processes. Shifting trends in the organization of production show the growing importance of transnational corporations (TNCs), and an increased "stretching of corporate activity and business networks across the world's major economic regions." They also demonstrate the prominence of TNCs in "organizing and managing cross-border business activity through the ownership of plants, outlets or subsidiaries in different countries."[32] Transnational corporations have been able to rapidly move both jobs and productive complexes to the lowest-cost regions of the world. This ability—combined with the computer-aided capacity to break down the production process into numerous different components, and then to research, design, manufacture, integrate, fabricate, and harmonize these different components in different parts of the world—has led to significant changes both in production patterns and in migration streams and migrant ways of life.

Contemporary global production has given transnational corporations, small and medium-sized enterprises, and their managerial elite far greater latitude in the establishment of different sites of productive activity *and* reproductive leisure. While the structural division of labor in the production of particular commodities has been well documented, the new types of divisions between production and *re*production have not received as much attention.[33] Both laborers and the owners and executives of businesses and corporations have often been required to move vast distances between the site of home and family, and the site of production. This kind of transnational movement of labor, including managerial labor, is an essential component of the globalization of production. There is relatively little choice for most manual and domestic laborers, who must move between nations polarized by differences in wealth and employment opportunities, and whose sites of social reproduction remain the cities in which they were born and where the bulk of their families remain. But for most entrepreneurs and executives, there is choice on both fronts: in the site of investment and in the site of family life. As Beck notes vis-à-vis transnational corporate choice: "In the manufactured and controlled jungle of global production, they are able *to decide for themselves their investment site, production site, tax site and residence site,* and to play these off against one another. As a result, top executives can live where it is nicest to live, and pay taxes where it is cheapest."[34]

The movements of both laborers and owner/executives between sites of production and social reproduction across national borders caused scholars and politicians in the early 1990s to "reconceptualize the nature of immigration, and to create a new vocabulary and theory to describe the transnational connections of contemporary immigrants."[35] They began to refer to the new immigrants as transmigrants, and to their lives as transnational. With the term "transnational," migration scholars were indicating a dynamic of migration and of migrant lives that was multi-local, often associated with a separation between the realms of production and social reproduction.

Transmigrant laborers generally work and settle in highly industrialized countries such as the United States, Canada, Germany, England, or Denmark; at the same time, however, they travel "home" frequently, communicate with those left behind on a weekly or daily basis, buy property and build houses and schools there, and send money and gifts back to family members.[36] They thus can be conceptualized as living within a network or "social field" that extends across national borders, and in which their actions and thoughts can be seen to be *simulta-*

*neously* connected and involved with more than one nation at once.[37] While producing goods or services in the advanced economies of the world, transmigrant laborers remain connected to their less developed "homeland" through various reproductive functions, often continuing to raise and educate children there, take care of elderly parents, receive medical attention and recover from illness and stress. They also often expect to retire to or die in their country of origin. For many transmigrants, this relationship is not just material in terms of these personal ties, but extends to a deeply felt nationalism that revolves around a desire to obtain justice for the homeland, which has been shut out of the cornucopia of global wealth and its concomitant displays of dignity and respect.[38]

The technological innovations of jet planes, electronic banking, television, email, home videos, and especially transatlantic phone service facilitated this form of simultaneous living; as physical distances were increasingly collapsed, migrants were able to participate more fully in the social, political and economic activities of more than one national site. In numerous empirical case studies, migrants have been shown to live fully and actively in two nations, and often to conduct business, politics, and family affairs in both.[39]

But while this growing body of research has aided our understanding of the transnational lives of migrant laborers, there is less scholarly output on the movements of transnational entrepreneurs and executives.[40] This lack of attention is especially significant because of the important question of *choice* in sites of production and reproduction. Members of the transnational laboring class have relatively little choice with respect to the site of either their productive activity or their reproductive "home." This is not the case for the transnational elite, however, who are often able to strategically manipulate or evade the regulatory systems of state borders and systems of governance, and who also are, more often than not, able to purchase citizenship in more than one locale.[41] With this increased flexibility, the transnational elite can choose their site of investment *and* the site in which they intend to raise a family, get an education for their children, rest, receive medical care, and retire. I believe that the spatial splitting of production and social reproduction and the question of choice in this area have great ramifications for the social and political, as well as the economic, realms of sending and receiving societies.

In the last two to three decades, especially in the economies of greater China, the internationalization of production has been greatly augmented by the incorporation of small and medium-sized enterprises that operate autonomously and within the larger TNC production and distribution networks.[42] Most transnational Chinese investors and entrepreneurs from Hong Kong choose productive sites for their factories in southern China.[43] For many of this group, one of the premier sites for social reproduction has been Canada. With its reputation for quality in the areas of education, health care, and the environment, and with a history of migration from southern China and Hong Kong, as well as the incentive of the Business Immigration Program, Canada became an extremely attractive destination for numerous wealthy business migrants from Hong Kong during the 1980s and 1990s.[44]

As I show throughout this book, the spatial splitting of production and reproduction was important in constituting a particular outlook and set of priorities for the migrants and their families, as well as in their general social and economic reception in Canada.[45] The business immigrants and their families were quite literally brought into the society on the basis of their economic "worth," and were expected to deliver on that value in their new home. As we shall see, however, their differing ways of understanding and being in the world, based in part on this economic worldview, were bitterly contested in the battles over the logic of liberalism as manifested in the production and reproduction of the Vancouver landscape and its associated institutions of governance.

## SITUATING NEOLIBERAL HEGEMONY IN GLOBAL SPACE

After nearly a decade of reading, teaching, and writing, I keep coming back to the question of hegemony. How does the social thinking of a particular historical bloc come to form the dominant mode of understanding how the world works, and how is this a fundamentally spatial question? More to the point, I am interested in how this social thinking can change. What are the processes by which new conceptions arise? How exactly does collective consciousness shift? Why do the events of particular historical moments bring about deep ruptures in dominant political thought and argument, and lead to new articulations and alliances?

In most contemporary discussions of hegemonic production there remains a primarily *abstract* use of the sites or "spaces" of society, and a general privileging of the historical over the spatial.[46] History clearly establishes the temporal layers on which subsequent hegemonic formations are established. But hegemonic formations are also made in the production of, movement through, and representation of space. The sites of spatial securing, the "spaces" of civil society or the state, are concrete, lived, and produced spaces. Hegemony is made not just in time, but also in space—in the taken-for-grantedness of bodies in specific places, what those bodies are, what those places are, and how they constitute and define each other. Every future moment of hegemonic formation is affected by the spaces of the past. And they are spaces produced and reproduced at different scales. This is a key point, because the widely assumed locus of hegemonic formation is in the institutional spaces of the state, but the idea of the fixed state as an implicit territorial container of these concrete spaces is an outmoded one. This political transformation has vast implications for how we need to rethink hegemony.

Althusser, for example, located hegemonic formation primarily in state institutions. He found the workings of ideology in the material, concrete practices of schools, family policy, laws, and so on.[47] Hall widened his theoretical catchment area to include the institutions of "civil society," arguing that authority won *outside* the state could be especially important for establishing legitimacy *for* the state.[48] Both these theorists, however, implicitly located state and civil society practices within "the nation." The primary story of the articulation of power in relation to the state was the story of institutional practices (associated with the state either directly or indirectly) contained within a physically defined national territory. This scholarly story made sense vis-à-vis the ideas and practices of most modern nation-states, where the nation was equated with a contained physical territory, and the idea of fixed containment was absolutely central to national narratives. But what if the state and civil society and even its citizens' practices extended beyond the national territory, as they do now? Globalization and transnational processes have produced major disjunctures for the state, where bodies moving outside the national territory must be "governed" by state practices in new ways. "The conception of a *national* community of citizens, made up of male breadwinners and female domestic workers, has been usurped by a new understanding in which not only are firms to be entrepreneurial, enterprising and innovative, but so too are political subjects."[49] These new political subjects are not just autonomous entrepreneurial actors, but are now fundamentally *unmoored* from previous discursive regimes relating to the physical territory of the nation-state.

What does this do to theories of hegemony? The first thing to consider is the distinction between the state and its practices, and the nation and its territory. Civil society, for the most part, has been unthinkingly equated with *national* society, with the institutions and practices of people located within the territorial borders of the nation-state. Stuart Hall's discussion of the rise of the neoliberal discourse of Thatcherism, for example, showed how numerous institutions of civil society helped to promote many of the supremely nationalist ideologies that coalesced to form the hegemonic bloc of Thatcherism—"nation before class, the organic unity of the English people, the coincidence between the 'English genius' and traditionalism," not to mention the profound media hype surrounding the Falklands/Malvinas War.[50] The question to ask is not whether the equation of civil society with national society worked, as Thatcherism clearly pervaded an entire decade, but whether this particular moment of suturing, this type of hegemonic articulation, could work again.

I believe that the hegemonic formation underpinning Thatcherism was able to succeed so dramatically because of the particular context in which state practices and national narratives could, for a time, coalesce. With the increasing imperative from business to transcend state borders, a transcendence materialized and visible in contemporary flows of money and migrants, the coalescence broke apart. This came to a particular dénouement at Maastricht, where the contradictions between the world of high finance, with its strong advocacy of closer links with the

European Union, and the world of nationalist, territorialist rhetoric (expressed most loudly during the Falklands/Malvinas War) became painfully apparent and led ultimately to Prime Minister Thatcher's downfall.

The ways that power coalesces in relation to the state (with, against, and around it) have changed dramatically in the past three decades and are continuing to transform in relation to contemporary globalizing forces. Cities, for example, are now positioned in a different functional role vis-à-vis both the state and the global economy. In the neoliberal vision, cities, as Jessop noted, are now perceived as key "engines of economic growth," as well as centers of innovation and crucial galvanizing forces of entrepreneurialism and competition. They are also envisaged as "managing the interface between the local economy and global flows, between the potentially conflicting demands of local well-being and those of international competitiveness, and between the challenges of social exclusion and global polarization and the continuing demands of liberalization, deregulation, privatization, and so on."[51] The spaces of cities are the spaces where the hegemonic struggles over liberalism are now being fought. Whose liberalism? Whose hegemony? The socially revisionist liberalism of "well-being" or the neoliberal mantra of "international competitiveness"? There are clearly new alliances, new struggles, new forms of subject formation, new forms of consciousness, new narratives, and new and ongoing imperatives to rework the ever-shifting articulations of state and nation, and nation and city. The disjunctures between neoliberal state practices and national or urban narratives of social liberalism, for example, create a rent in the fabric, a tear in the sutures, a moment where the taken-for-granted becomes suddenly visible.

But the question remains, When do people understand and act on these visible disjunctures and contradictions? How does the collective thinking of a society *shift*? This, as Hall succinctly put it, is "the problem of understanding how already positioned subjects can be effectively detached from their points of application and effectively *repositioned* by a new set of discourses."[52] It is the problem of thinking about change, the transformation of both subjectivity and society. The study of hegemony begins, for me, with the understanding of change as not just possible, but constitutive of hegemonic formations. Indeed, not only is the process one of change and of tendencies rather than of final moments, but also it is clearly not a linear process, nor is the formation of counter-hegemonies necessarily progressive. Hegemony is not so easily produced or maintained by conservative forces as its theoretical legacy seems to bear witness; the acquiescence of dominated groups to their own subordination is always uneven, fragmented, and partial.

How then shall we theorize hegemony, if we take as a starting point its fundamentally unessential character, its processual nature, its dialectical constitution and shifting, chameleonlike qualities? The concept loses coherence without a sense of domination and subordination, a domination whereby an elite group wins and maintains power not by force but through the acquiescence of the subordinated group. This acquiescence is based primarily on the insistent and inexorable effects of numerous unequal institutional apparatuses in society that take on the characteristic of the self-evident, the normal, and the correct. Further, these institutions, and the unequal values they embody and uphold, become a concrete, material part of the social world inhabited by both the dominant and subordinate, and thus become part of the everyday practices and lifeworld experiences of these groups in a mutually reaffirming cycle. Hegemony thus can be thought of as an interconnected system or formation where "specific distributions of power and interest" are related to "the whole social process."[53]

Most contemporary theorists would not quibble with this as a starting point, but there is considerable disagreement as to exactly how this acquiescence works. It seems to me that much of this disagreement arises from the tensions between theorizations that rely primarily on abstractions, and those that apply to particular cases. Even within the corpus of Marx's work there is a tension between the early pronouncements of *The German Ideology* and the empirically detailed analysis of *The Eighteenth Brumaire*. As examples, here are a few sentences from the former work: "As individuals express their life, so they are. What they are, therefore, coincides with their production, both with *what* they produce and with *how* they produce. The nature of individuals thus depends on the material conditions determining their production"; "the production of ideas, of conceptions, of consciousness, is at first directly interwoven with the material activity and the material intercourse of men, the language of real life. Conceiving, thinking, the mental intercourse of men, appear at this stage as the direct efflux of their material behavior." And most definitively and famously: "The ideas of the ruling class are in every epoch the ruling ideas, i.e. the class which is the ruling *material* force of society, is at the same time its ruling *intellectual* force. The class which has the means of material production at its disposal, has control at the same time over the means of mental production, so that thereby, generally speaking, the ideas of those who lack the means of mental production are subject to it."[54] These are, of course, the type of quotes taken up by those who wish to stress the scientific and analytical logic and predictive ability of Marxist thought.

However, if you compare these with any of several passages from *The Eighteenth Brumaire*, it is possible to see Marx's historical passion, which spills forth and releases the abstracted and static "peasants" and "bourgeoisie" into real people struggling with internally divided classes, conflicting desires, historical aberrations and personages, unintended consequences, and wildly oscillating victories and defeats within a space of just three years.[55] In *Brumaire,* one is given insight into the endless internal struggles and contradictions of history, the errors of calculation, and the tendencies never fully realized. As Gramsci points out, reading this type of historical work by Marx (the other major example is *The Civil War in France*) serves as a constant precaution against identifying structures statically or ahistorically, which Marx himself avoided in his own empirical work.[56] Other Marxist scholars came to the same conclusions, including Luxembourg, who wrote of the "impossibility of treating certain questions of the philosophy of praxis in so far as they have not yet become *actual* for the course of history in general or that of a given social grouping."[57]

In the following discussion of the partial hegemony of social liberalism in the mid twentieth century, and of the unraveling and reconstitution of liberalism as currently struggled over in urban space, I am mindful of the caveats of these theorists of hegemony. It is "in the course of history" *and in space* that one is able to tease out the contradictions of liberalism and neoliberalism both, and to see how different actors draw on the dominant narratives of both philosophies in the struggle to supplant or sustain new narratives and meanings. Actually existing liberalism and neoliberalism are constructed, contradictory, and profoundly contestable processes with no linear pasts and no linear futures.[58] And despite their identical provenance in seventeenth-century British thought, these two terms have come to mean vastly different things to most people.

As I have been arguing all along, context counts, and liberalism has a history and a geography to it. In most cases, the historical path taken in the twentieth century has been, as Polanyi put it, the path of an increasing embeddedness of "market economies" in "market societies." He was referring by this to the reregulation and embedding of capitalism within various apparatuses of control and oversight, most importantly, the state.[59] The development of a general consensus on Keynesian ideas concerning the importance of planning, full employment, and the maintenance of a minimum level of social reproduction represented both an economic *and* a social philosophy about the organization of society. The national welfare state simply could not have developed as it did without a complementary hegemonic shift in liberal thinking—away from an emphasis on private property and an obsession about the fundamentally predatory and tyrannical nature of the state, and toward a far greater emphasis on the social and economic foundations of inequality and the potentially positive role of the state as "an organized expression of organic community."[60]

Indeed, the idea of social citizenship, the foundational belief that all members of the national society have a right to participate in the democratic polity, and that this is only possible given a minimum standard of social reproduction, became taken for granted in many societies, perhaps especially in Canada.[61] That it was the responsibility of the state to ensure this minimum level of reproduction became a normative way of thinking in the twentieth century, and most citizens of the nation assumed a degree of entitlement for all Canadians in the arenas of health care, housing, education, and workplace protections and benefits. The expectations regarding these material entitlements were matched with socially liberal assumptions concerning the "rights" of all to respect, dignity, and tolerance by members of the national community, despite any individual deviance from "normal" Canadian society and culture. Indeed, in the heyday of social liberalism in the 1960s and 1970s, difference was promoted as not just necessary, but actually positive for national objectives of unity and coherence.

Numerous scholars have focused on the rise and seeming consolidation of the new hegemony of neoliberalism, but few have unpacked the simultaneous unraveling of the ideological tenets of social liberalism.[62] And fewer still have analyzed *the connections* between these two processes, which are contested and take shape unevenly across space. Neoliberal hegemony might be waxing at the scale of the state and the globe, but at the urban scale, where different forms of liberalism hold another type of hegemonic sway, the game is far from over. As Jessop wrote: "Even where both the national and international levels are dominated by attempts to promote a neoliberal regime shift, the urban level may be characterized more by neocorporatism, neostatism, and neocommunitarianism. Indeed this last pattern is particularly linked to attempts to manage issues of social exclusion and social cohesion at the urban level even in the most strongly neoliberal cases."[63]

## NEO/LIBERALISM

The "neo" in neoliberalism is confusing because it seems to indicate a new and improved version of the commodity that has been on display in philosophy shops for the past several centuries. Is this an "instant replay" of liberalism, possibly with a faster rewind mechanism or a sharper lens? Is it simply the latest model of Lockean principles of individual liberty; private property; freedom of speech, association, and contract; and the all-important limitations on the authority of the state? Or is it something qualitatively different because of its history, its location in time *after* the postwar Keynesian moment? What does the "neo" in neoliberalism do to our understanding of liberalism and vice versa?

Agnew and Corbridge described a hegemonic "transnational liberalism" slouching toward world domination, and Katznelson wrote that "liberalism today has become a meta-ideology." Both texts seemed to bracket liberalism and neoliberalism as one and the same.[64] But Katznelson also noted that there are contending kinds of liberalism, and that, despite their seemingly current hegemonic status, they all rest on a fragile basis of moral and instrumental political reason. One of the tensions I want to investigate here is the relationship between liberalism and neoliberalism on the foundations of this shared basis of abstract political reason, and the differences that have developed between them as a result of the contingent histories and geographies of actually existing neo/liberalism(s) in Canada.

Owing primarily to Canada's political history as a former British colony (outside of Québec), the connections with British culture and a British liberal lineage remain predominant over other variants of liberalism in most of the country.[65] What are and have been the primary features of a seventeenth-century legacy of British liberal thought as it is developed in Canadian self-conceptualizations and systems of government and governmentality? What are the "keywords" of liberalism that are important to know in order to understand the types of narrative familiar to most Canadians, and that have become the normative background for social intercourse? What is the philosophical taken for granted, and how are these normative assumptions either augmented or challenged by neoliberal practice and discourse?

Early British liberalism emerged with the writings of John Locke and was developed further through the works of Jeremy Bentham, James Mill, John Stuart Mill, Herbert Spencer, Sir Henry Maine, and numerous others. Although there were as many varieties of liberal thought in the early years of its development as there are in contemporary discourse, the basic "package" of early liberal tenets rested on a commitment to individual liberty and dignity, the guarantee of individual rights of property, the freedom of expression and association, the right to representative institutions constituted within a democratic framework, and the right and responsibility to limit the authority of the state.[66] These tenets formed a general conception of persons and society distinct from the previous epoch and were distinctively "modern" in character. John Gray categorized the key elements of this conception as "individualist" with respect to the moral primacy of the person, "universalist" through its conferral of the same moral status on all, "egalitarian" in terms of its affirmation of the "moral unity of the human species," and "meliorist" in the sense of the corrigibility of political and social arrangements.[67] While different aspects of these tenets have been privileged by different thinkers, the general claims that relate to the autonomy of the individual and the commitment to liberty and universality are upheld by all those working within this philosophical tradition.

Critics of liberalism from outside its fold have attacked its premises from a variety of positions. Most of these critique liberalism on the basis of its inherent system of exclusions. In a well-known critique based on the exclusions of class, *The Political Theory of Possessive Individualism*, C. B. Macpherson showed the theoretical inconsistencies in Locke's promulgation of universal rationality. While arguing for the possibility of universal political rights based on the ideal of universal rationality, Locke clearly relied on an implicit historical assumption of "differential" rationality based on class position. The assumption of superiority of intellectual rationality based on the ownership of property was so naturalized in the everyday sphere of social life that Locke was able to take for granted the understanding that his idea of "universality" was necessarily bracketed by class. Further, this normalized differential was crucial in facilitating the growth and expansion of capitalism in Britain. Macpherson wrote of this form of political exclusion: "[Locke] justifies, as natural, a class differential in rights and in rationality, and by doing so provides a moral basis for capitalist society."[68]

Similarly, in *The Sexual Contract*, Carole Pateman explicated liberalism's fundamentally sexist exclusions and implications. She focused on the ways in which the universal "freedom" implied by the concept of the social contract in liberal theory relied, in practice, on the subjugation of women through the sexual contract. Although framed as a move to a civil society fundamentally opposed to patriarchy, the social contract actually established and upheld men's domination and political rights over women. She showed further how Locke's assumptions about the "natural" dominance and political rights of husbands over their wives through the conjugal bond effectively undermined the possibility of women's owning status as individual agents. They were thus deprived of their political rights as rational individuals and excluded from full participation in the production of civil society. Pateman described Lockean liberal formulations on this subject: "We know that wives should be subject, Locke writes, because 'generally the Laws of mankind and customs of Nations have ordered it so; *and there is, I grant, a Foundation in Nature for it.*' The foundation in nature that ensures that the will of the husband and not that of the wife prevails is that the husband is 'the abler and the stronger.' Women, that is to say, are not free and equal 'individuals' but natural subjects."[69] Thus in both of these examples, the "universal" subject of liberalism and the condition for universal rationality were shown to be false; both women and propertyless men were effectively excluded from participation in or production of civil society as a result of existing patterns of class and gender domination and subjugation in British national society.

In a third example of exclusion, that of colonial subjects, Uday Mehta demonstrated the ways in which liberal exclusions not only occurred through the historical exigencies of patriarchy and capitalism, but also were immanent in Locke's texts themselves.[70] For liberals to be able to endorse empire as legitimate—despite the seeming paradox of imposing a lack of liberty, dignity, and right to self-ownership on "subject" populations in the colonies— it was necessary to develop certain theoretical strategies for delimiting the rights of universal membership. Mehta argued that Locke and other British thinkers in the liberal tradition relied on the sociocultural conventions of their social world to limit membership in the political circle. Colonial subjects were unable to consent to join because they lacked reason, that is, the reasoning needed to distinguish between various kinds of social norms and hierarchical distinctions. While abstract human nature was made to seem equal in Locke's writings, vindicating his claims of a common humanity and of universal possibility and opportunity, actual human beings were circumscribed from these possibilities owing to their "irrational" inability to grasp the all-important social distinctions of a class-based society. Through systems of education and through inhabiting a particular social world, class privilege was inscribed in the minds, on the bodies, and in the landscape of the gentry, bringing with it the distinction and breed-

ing necessary for understanding the dictates of "reason," and thus
for inclusion in civil society. Clearly, in the Lockean worldview,
colonial subjects did not possess the capacity to comprehend
these types of subtle distinctions and thus could not expect to be
full participants in political society. As Mehta pointed out:

> Terms such as "English gentry," "breeding," "gentleman," "honor,"
> "discretion," "inheritance," and "servant" derive their meaning and
> significance from a specific set of cultural norms. They refer to a con-
> stellation of social practices, riddled with a hierarchical and exclu-
> sionary density. They draw on and encourage conceptions of human
> beings that are far from abstract and universal, and in which the
> anthropological minimum is buried under a thick set of social inscrip-
> tions and signals. They chart a terrain full of social credentials. It is a
> terrain that the natural individual, equipped with universal capacities,
> must negotiate before these capacities assume the form necessary for
> political inclusion. In this, they circumscribe and order the particular
> form that the universalistic foundations of Lockean liberalism assume.
> It is a form that can and historically has left an exclusionary imprint
> in the concrete instantiation of liberal practices.[71]

I will return to a discussion of this exclusionary impulse at the
heart of liberal theory. But first I want to outline some of the ways
in which these tenets of liberal theory infected political institutions and the political orientations of British and British-Canadian
subjects. It is important to note that, despite the inherent and fundamental exclusions in liberal philosophy and practice in the seventeenth and eighteenth centuries, the professed liberal emphasis on universality and inclusion became a reality for greater
numbers of people in the centuries that followed. Functional liberalism in Britain expanded greatly in terms of the enfranchisement of propertyless men and then women and the reformation
of the legal system in the nineteenth and twentieth centuries.
With respect to the demand for full political participation, T. H.
Marshall characterized this era as one of expanding citizenship,
with systematic gains in the arena of political citizenship in the
nineteenth century and social citizenship in the twentieth.[72] For
midcentury liberals such as Marshall in Britain and John Dewey
in the United States, these gains were integral to the great promise of liberalism.

The liberal conviction of theorists such as these rested on
the steady and linear increase in economic, political, and social

enfranchisement of all members of the nation. Further, part of their inclusionary crusade was founded on the belief that greater enfranchisement was essential to the successful development of the nation. *Nationalism*, the augmentation of national identity and pride in the munificent and tolerant national community, was thus an inherent component of revisionist liberals who sought to ameliorate classical liberalism and purge it of its worst exclusionary excesses. Marshall's famous essay on social citizenship was based on his positive perception of Britain's historical trajectory from the vantage point of the mid twentieth century;[73] Dewey's philosophical promulgations were based on the idea of the limitless promise of his beloved American nation and the successful exportation of liberalism, American style, around the world.[74]

This idea of progress toward a more just and equitable society full of tolerance and opportunity for all was a fundamental aspect of one key wing of nineteenth- and twentieth-century liberal thought.[75] Indeed, there was a profound sense of betrayal among many liberals in the early twentieth century that this promise of tolerance and inclusion was not adequately fulfilled, and in fact seemed to be declining because of class polarization and the growing comfort of the middle classes, who had been among the most important early agitators for more social and political inclusiveness within the liberal national fold.[76]

The uneven development and class polarization caused by an unfettered market economy were seen to be inhibiting the promise of both liberalism's ultimate utopian vision and the imagined community of the nation. Numerous public intellectuals, including Harold Laski, made a strong push at this time for a more insurgent variety of liberalism. They argued persuasively for a redistributive component to be developed further within liberal theory to combat ongoing capitalist inequities. This redistributive tradition within British liberalism grew in the context of the expansion of democratic institutions, and a succession of capitalist accumulation crises of the nineteenth and early twentieth centuries. It became an essential feature of the debates over the principles of justice and fairness that were the hallmark of late-twentieth-century liberalism.[77]

Liberalism has *always* aspired (in its own self-conception) to articulate the political and social norms of tolerance and comity.[78] But only in the second half of the twentieth century were many of these aspirations realized in more than a token fashion. Contemporary liberal theorists like John Rawls were concerned with two related components: "the commitment to the freedom of the individual embodied in the standard liberal support for civil liberties, and that belief in equality of opportunity and a more egalitarian distribution of resources than would result from the market alone which leads to support for a redistributive welfare state."[79] The latter component is the one farthest removed from classical or Lockean liberal thinking and in fact has been attacked as a kind of socialist liberalism by libertarian liberals such as Robert Nozick. Nozick's position holds that the redistributive element in Rawlsian liberal theory violates the first, or "civil rights," component of the liberal "package"—the individual's rights to self-ownership and to property—and hence is fundamentally antiliberal in its primary set of propositions.[80]

During the twentieth century, most modern liberal states in the West adopted the liberal revisionist principles of equality of opportunity, state intervention and support, and redistribution (to varying degrees) as narratives of state legitimacy.[81] Further, this variegated set of ideas became increasingly important with respect to the general societal "consent" to the state's authority to govern.[82] The rise of welfarism became part of an assumed obligation to create opportunity and a minimum level of support in the search for a plural and tolerant national community. The provision of social services by the state was one aspect of a compact of socioeconomic justice for which unions and numerous other social forces had struggled for more than a century. These social forces included many revisionist liberal thinkers who were influenced by the democratic transformations in society, and whose ideas worked back on society as they began to articulate commonsense understandings of state-society relations and the norms of obligation, fairness, and entitlement.[83]

Through the nineteenth and twentieth centuries, the increasing political enfranchisement of different groups and concomitant demand for economic opportunity and equity was reflected in

the revisionist liberal theory promulgated by midcentury scholars such as Dewey in the United States and Laski and Marshall
in Britain. Critiques of classical or economic liberalism with a
strict Lockean emphasis on private property and state limitation
came from this redistributive wing *within* liberalism. This group
of thinkers wanted to hold onto the classically liberal tenets of
individual choice, autonomy, liberty, and free inquiry, separate
from the overweening emphasis on private property, contract,
and an obsession with the tyranny of the state.[84] It is this socially
progressive wing of liberalism—what I term "social liberalism"—
that became partially hegemonic in philosophy and practice in
Canada, as elsewhere, reflecting the rise of the welfare state and
also aiding in its philosophical justification. It is also this social
liberalism that confronts and has been confronted most directly
by the current proponents of neoliberalism in Canada.[85]

Despite great advances in welfare entitlements in Canada
through the early and mid twentieth century, the liberal goals of
equity and inclusion for all remained elusive. The various forms
of injustice and exclusion that remained, however, were not just
accidents of poor implementation, but were integral to the constitution of liberalism. As I discussed earlier, an exclusionary
impulse is located at the heart of British liberalism, a political heritage that has continued to influence Canadian political culture.
Despite revisionist efforts to ameliorate the inequities caused by
capitalism, the central "civil rights" tenets of liberalism—individualism, individual choice, dignity, freedom and rationality—
are premised on a form of reason and of rational behavior which
is culturally inscribed and can *never* be completely accessible to
the outsider. "Reason" and "normality" are often used either
explicitly or implicitly to limit claims of universal membership
in the political society. Further, they are concepts that are deeply
linked with spatial practices and assumptions, as we shall see.[86]

Many liberal critics have noted how the concept of reason is
tightly linked with the *familiar* in liberal theory. For most liberal thinkers, the unfamiliar can be either assimilated into the
generalizable structures of current (rational) thought, or shunned
as childish, superstitious, or "irrational."[87] Within liberal thought
there is great acceptance of and tolerance for differences that operate within the parameters of established structures of generality,

but little if any room for a difference that is fundamental, that challenges the nature of reason itself. Mehta writes of the early liberal thinkers:

> The historian in James Mill, the legislator in Bentham, the educator in Macaulay, and the apostle of progress and individuality in J.S. Mill, all, I believe, fail in the challenge posed by the unfamiliar; because when faced with it they do no more than "repeat," presume on, and assert (this is where power becomes relevant) the familiar structures of the generalities that inform the reasonable, the useful, the knowledgeable and the progressive. These generalities constitute the ground of a cosmopolitanism because in a single glance and without having *experienced* any of it, they make it possible to compare and classify the world. But that glance is braided with the urge to dominate the world, because the language of those comparisons is not neutral and cannot avoid notions of superiority and inferiority, backward and progressive, and higher and lower. Urges can of course be resisted, and liberals offer ample evidence of this ability, which is why I do not claim that liberalism *must* be imperialistic, only that the urge is *internal* to it.[88]

This intolerance of the unfamiliar had great implications for the imperial project, in which a different framing of experience by colonial subjects was deemed irrational and in need of paternalistic education and reform by colonial masters. It provided the justification and the desire to conquer and control other areas of the world deemed backward and uncivilized; it provided a cover for empire. I believe that the inability of liberalism to engage deeply with the unfamiliar is also at the heart of the struggles over transnational migration and immigrant challenges to the spatial practices and norms of Canadian society.

The wealthy transnational migrants from Hong Kong were unfamiliar agents in Vancouver, operating with a different, nonnational viewpoint and *neo*liberal set of assumptions about what constituted reason and the rational. Yet they had the *power* (unlike former colonial subjects) to defend their own worldviews. They also had the power to be imperialistic, to advance their own understandings and rationalities in spatial terms. Further, other individuals and institutions in the city and the region were able to ally with the immigrants and manipulate the differing understandings and inherent contradictions within social liberalism to gain economic or political advantage for their own projects. This strategy was efficacious in many cases as a result of the context in which liberalism had been practiced in Canada: the actual limits to universal inclusion based on the historical circumstances of racism and colonialism as they had played out through time and space in Vancouver. It is the fundamental illiberalism of liberalism at its core—the inability to tolerate the truly unfamiliar or the thick experiences of another way of life and mode of being "at home"—that was strategically manipulated by the alliance of wealthy immigrants, developers, and politicians in their challenge to social liberalism and in the advocacy of neoliberal policy in the city.[89]

The challenge to social liberalism in urban policy and in the national narratives of multicultural tolerance in Canada thus emerged from a particular, historically specific alliance that upheld an economically liberal and global orientation, one where property rights took precedence. This economic liberalism was, in many respects, a return to an earlier, Lockean set of principles, where property rights and the abrogation of state authority comprised the core philosophical concepts. But it also indicates the complex and hybrid nature of new political configurations, and the ways in which multivocal social movements derive political purchase through drawing on different strands of "commonsense" understandings in society.

While it looks like a "return" to classical, Lockean liberalism, and could thus possibly be theorized as a kind of "transnational liberalism" if viewed from above, the particular configurations of power and the specific historical and geographical circumstances of this struggle made it something quite different. The development of the form of social liberalism I have described here is particularly *national* in orientation and nationalistic in flavor, and was ruptured in terms of its hegemonic primacy at the urban scale through the arrival of unfamiliar, *trans*national agents. Actually existing liberalism, I would argue, is a fundamentally national formation; it is neoliberalism that has global ambitions.

Neoliberalism thus can be read as both an ideological rationalization for free-market globalization *and* an ideological force in opposition to a nationally based agenda of social liberalism. It is a process of state restructuring in which the Keynesian reforms of the twentieth century are rolled back through various processes of deregulation and dismantlement, and the smooth, spaceless functioning of a laissez-faire marketplace is rolled out through regulatory reform and active state intervention. In the roll-back phase of the 1980s, as Peck and Tickell noted, the welfarist principles of redistribution were undermined and discredited and many publicly owned goods and services were privatized.[90] In the following decade, neoliberal regimes continued to sell newly privatized goods and services, especially on the urban scale, as well as *the idea of* perpetual marketing and reform—the complete and ongoing mobilization and commodification of the city, urban life, and even the individuals acting therein.[91]

Contemporary scholars of governmentality such as Nikolas Rose examine the ways in which neoliberal changes such as these have usurped the hegemonic discourse of welfarism and become entrenched through the disciplining of numerous institutions and realms in society, including schools, hospitals, the workplace, and government agencies.[92] The concept of "freedom," for example, a discourse of market and of self, has infected social life to the point where all government becomes self-government; citizens have come to understand that they must not "devolve responsibilities for health, welfare, security and mutual care upon 'the state,' but take responsibility for their own conduct and its consequence in the name of their own self-realization. The well-being of all, that is to say, has increasingly come to be seen as a consequence of the responsible self-government of each, and the demands of freedom have become even more closely intertwined with the government of subjectivity."[93]

This type of neo-Foucauldian work investigates the constitution of the neoliberal subject rather than focusing primarily on neoliberal policy. It also brings to light the multiple microtechnologies of power that are involved in the reformulation of rule and the reorganization of society. As such, it is able to avoid the kinds of generalizing tendencies of works on neoliberalism that focus primarily on shifts in state doctrines and political projects conceived and implemented from above. Nevertheless, as Larner and others have noted, the governmentality literature, in general, tends to eschew the "messy actualities" of particular projects and programs, and to focus instead on broad themes and theoretical arguments. Shunning the kind of detailed ethnographic work that usually evokes individual and group organizing, community building, humor, sarcasm, love, resentment, and resistance *alongside* institutional, political, bureaucratic, and governmental discourses has led to a "body of work [which] privileges official discourses, with the result that it is difficult to recognize the imbrication of resistance and rule."[94]

My point here is that history and geography are always key both to the development of certain kinds of thought and ways of being in the world, and to the ways in which philosophical frameworks are produced and implemented in both policy and practice. But a key corollary to this is the understanding that the conceptual apparatus in which we locate the contingent is itself changing. Not only is it necessary to make explicit the genealogies of time and space which infect the narratives of liberalism and neoliberalism, but also it is imperative to confront the overwhelming power of the nation format within which so much theoretical work is positioned. The spaces of social as well as economic life have undergone a transformation in the last several decades. These spaces are the streets on which people walk, the houses they build, the trees they chop down, and the schools in which their children are educated; they are not the metaphorical spaces of abstract institutions. To discern the shifting experiences of self and society and the ways in which liberalism and neoliberalism are ebbing, flowing, and overlapping, we must consider how these hegemonic formations and our theories of them are imbricated in the new global spaces of everyday life.

The disjunctures between a neoliberal rhetoric of neutral space and economic competition, and a national liberal rhetoric of pride of place, universal inclusion, and social harmony, became visible and contestable in Vancouver through the local urban struggles over trees, houses, zoning and multicultural ideology. In the course of these struggles, where different actors and alliances drew on different strands of liberal thought, it became apparent that the rhetorical tools of social liberalism often employed against neoliberal practices and rhetoric were flawed. The exclusionary and imperialist origins of liberal thought made the spirited defense of the "social citizenship" of mid-twentieth-century welfarism in Canada deeply problematic.

The cracked foundations on which many liberal assumptions rested became particularly manifest in the highly racialized struggles over landscape, houses, zoning, and ideologies of togetherness, when unfamiliar outsiders from the global ecumene disturbed the nationalist fantasies of equal entitlement, tolerance, unity, and coherence. Utilizing a critique of state-led liberalism based on its history of racial exclusion evident in the landscape, neoliberal proponents of change in Vancouver were able to shore up their rhetorical defenses from socially liberal attacks. In this vein, the neoliberal discourse of rational market forces and the "regeneration" of the city were simultaneously heralded as freedom from the conflictual scars of racial antagonisms as they had played out historically in Vancouver's urban milieu.

Cities have become the key sites of contemporary struggle over neoliberal hegemony, not just because they are the "most visibly denuded victims of roll-back neoliberalism,"[95] although that is certainly the case, but even more because they have social histories and geographies that provide actually existing liberalism with both its greatest hopes and its deepest problems. This work centers on the history and geography of one of these cities. By locating liberalism and neoliberalism in the everyday life of late-twentieth-century Vancouver, I hope to show how liberal formations are produced, not just as policies or rationalities from the state or from the philosophical propositions of intellectuals, but also from "a shout in the street."[96] Liberal formations are made in the actions and narratives of individuals and groups in public hearings on land use, demonstrations against tree removal, letters to the editor, the picketing of "monster" houses, and a number of other forms of social mobilization.

The book is structured as follows: each chapter emphasizes an urban struggle or set of struggles around which a number of discourses have crystallized. In the second chapter I introduce the marketing of Vancouver on the global stage and the ensuing widespread gentrification of working-class and middle-class neighborhoods. In a now classic pattern of disinvestment, reinvestment, and displacement, huge swaths of downtown property were remade in a decade into sparkling new corporate and residential towers.[97] In addition, large sectors of inner suburban residential housing were razed and rebuilt to meet the demands of a new dominant class arriving from Hong Kong and, to a lesser extent, Taiwan. Although the process followed the usual displacement of working-class residents from the single-residency hotels and cheap apartment buildings of the city center, and involved the types of public-private partnerships now well documented in cities across the globe, two other features of this transformation indicated an even greater alteration in the relationship between capital and the state and between state and society. These features, in particular, became the focus of residential struggle and the subject of discursive disputes vis-à-vis the intentions and meanings of Canadian liberal policy in the urban sphere.

The first neoliberal feature of urban and provincial governance that caught the public's eye was the striking boldness with which politicians, as well as real-estate capitalists, marketed the city overseas. Not only was the neoliberal state clearly an agent rather than a regulator of the market, "becoming, in effect, a junior if highly active partner to global capital,"[98] but also it was clearly targeting a specific region of the world: the Pacific Rim. The second feature of contemporary urban governance that caused dissent was the clear movement away from any substantive interest in social reproduction in the city, and toward a productive economy *based on real-estate development*. Smith writes of this feature of global, "third-wave" gentrification: "Most crucially, real-estate development becomes a centerpiece of the city's *productive* economy, an end in itself, justified by appeals to jobs, taxes, and tourism. In ways that could hardly have been envisaged in the 1960s, the construction of new gentrification complexes in central cities across the world has become an increasingly unassailable capital accumulation strategy for competing urban economies."[99] These two features of neoliberal urban policy were directly linked with one particularly fraught transformation of the city. This was the displacement of white middle-class women from their apartments and houses, followed by the construction of new luxury condominiums and large "monster" houses primarily developed and marketed for Hong Kong and Taiwanese Chinese buyers. This displacement had both racial and class implications. The ensuing struggle over these gentrified spaces became a struggle over both the language of race and class and over state-society relations and neoliberal urban policy.

The third chapter examines the discourse of multiculturalism. I investigate the ways in which the liberal ideology of multiculturalism was frayed and undermined because of its clear manipulation by neoliberal state actors. In Canada, the language of multiculturalism is an important rhetoric of national cohesion and of state legitimacy in its ability to "control" difference. Tolerance is assumed to be a rational feature of contemporary democratic liberalism, and an important building block of cohesion and acceptance of difference for the good of the "national" community. With the cynical manipulation of the ideology of tolerance by politicians and capitalists intent on capitalizing on Vancouver's integration into the global property market, however, the underlying values of liberalism became questioned and questionable for the city's residents. Further, what soon became obvious, even to Canadians outside Québec, was the domination of one branch of liberalism over another. Despite the attempt by federal politicians to appear neutral and accepting of difference, the conflicts over liberal multiculturalism quickly elucidated the strong privileging of a British procedural version of liberalism over the more communitarian variety of Québécois cultural nationalism.[100] The inherent and insurmountable contradictions between procedural liberalism and a neocommunitarian variation heralded by Québec and trumpeted by other groups seeking to protect cultural difference created a crack in the liberal foundations on which the narrative of multiculturalism "for" the nation rested.

In the fourth chapter I look at the ways in which the inner workings of urban-land governance in Vancouver were challenged by both long-term residents and the transnational immigrants. Drawing from three cases in which the hegemonic discourse of land planning and dispensation was either upheld or subverted by these groups and their various alliances, I indicate the multiple ways in which popular interpretations of Canadian liberalism were used to promote specific political and economic agendas related to the Vancouver landscape. In all three cases, complex brews of communitarianism, social liberalism, and neoliberalism were heralded by different actors and institutions as the most rational, normal, or obvious answer to the problems besetting the city. Older white residents relied on implicit assumptions about the community good based on a British-inflected cultural vision of stability and harmony; neoliberal academics, planners, and politicians involved with land governance, along with several of the transnational immigrants, countered these dominant values with alternative conceptualizations of the good life. In the former case, the vision was underpinned with a sense of territorial identification at the national scale; in the latter, the imagined community was diasporic and global.

In the final chapter I return to the earliest and most visceral reactions to neoliberal urban governance, the changing qualities of everyday domestic life. Through an investigation of the meanings of house and home in Vancouver, I take apart the ways in which middle-class white understandings of domesticity were threatened by a sharp decline in the standard of living brought about through neoliberal policy, particularly the declining interest in or financial support for systems of social reproduction in the city. Poor and working-class families and their housing were threatened to a much greater extent through the domino effect of property market and rental activity. But white middle-class anxiety captured the declining status of "white" Canadian capital vis-à-vis "Asian" or "global" capital in a way that produced a far greater disruption for liberal understandings of the rational, the normal, the familiar, and the tolerant.

Once again the issue of race dominated the widely publicized struggle over the housing and landscape changes in the city, and called forth both the exclusionary kernel at the heart of liberal thought and the difficulties inherent in contesting neoliberal urban policy with the flawed tools of social liberalism. The domestic sense of loss bitterly expressed by residents who had lost houses or felt a strong decline of neighborhood character elicited an era of racial privilege that evoked the colonialist mentality of white property ownership and Chinese positions of servility from the late nineteenth and early twentieth centuries. Domestic white privilege in west-side Vancouver, as many scholars have shown in other contexts, had been upheld through the processes of racialization that were sedimented in the landscape through time and came to be perceived as normal and natural.[101] Thus when fighting economic dislocation, residents called on this "natural" landscape as theirs by right, erasing the colonial legacy of racial and class segregation integral to its construction. This elision, however, was disallowed by the very presence of the wealthy Chinese home buyers who challenged notions of the normal, the natural, and the self-evident by their material practices and symbolic use of space. Further, these disruptions to white middle- and upper-class normative visions of appropriate domestic practices and behavior could not be effectively countered by long-term residents because of the strong privileging of economic rights to private property based in an ascendant neoliberal formulation of liberalism to which these wealthy residents normally ascribed.

In the long march of history, how is it possible to determine the hows and whys of hegemonic restructuring? What are the signals of paradigmatic change that literally stand out in the landscape, but need to be read with a critical eye to be seen? This book is an investigation of these signals, and of the ways that local regimes of power and knowledge shift in a complex dance with global economic forces. Making theoretical links between scales, and conceptualizing the hegemonic process as involving actual spatial struggles through time, allows us to interpret urban transformation as ultimately and inevitably linked with the broader discursive and material shifts occurring worldwide. This book charts one such transformational moment and shows how the processes of disjuncture and reorganization are always constituted in space—not the discursive, metaphoric spaces of the nation, but the actually existing spaces of homes, neighborhoods, planning rooms, provincial meetings, zoning hearings, and city council chambers.

# 2    Vancouver Goes Global

As a state strategy, urban neoliberalism creates new conditions for the accumulation of capital; yet it also inevitably creates more fissures in which urban resistance and social change can take root.
—Roger Keil, "Common-Sense Neoliberalism"

IN THE 1980s, urban policy in Vancouver began to shift from an ethic of social liberalism toward a philosophical and practical framework of neoliberalism. Neil Smith's concept of the "revanchist" city is useful here, as it indicates not just a move toward a new political framework, but also a deliberate and often brutal repudiation of the ethics and practices of the past.[1] The French word *revanche* means "revenge," and Smith uses the word quite literally in the context of an urban politics that exacts retribution on the liberal or socialist leanings of earlier regimes.[2] *Neo*liberalism in the city and in its global incarnations is often distinguished thus, as a strong ideological repudiation of the principles of social liberalism, and at the same time as a concerted effort to extend the reach of neoliberal programs. In this sense, one can see the underbelly of the beast, an impulse to clear out the "detritus" of society and history, the various blockages and messiness of poverty and racial and class antagonism—along with any attempts to alleviate them—and to render clear and free the so-called level playing field of the market.

In Vancouver municipal politics, a liberal reform party called TEAM (The Electors Action Movement) was elected in 1968 and maintained power through the late 1970s. At the provincial level, the New Democratic Party (NDP), a socially liberal provincial party, was elected in 1972 and remained in control during this decade. At the federal level, parliament was headed by Prime Minister Pierre Trudeau's Liberal government. Under Canada's constitutional system, provincial control over urban policy is extremely strong, and cities do not hold much autonomy. In this decade, however, all three levels of government were committed, to varying degrees, to an interventionist state role in providing urban services and increasing economic redistribution, and to implementing this ethic in the urban landscape. For example, a large mixed-housing development built on the south shore of False Creek during these years was designed to facilitate a "delightful and humane living environment," as well as to provide some social housing for poorer residents of the city. Ley wrote of this development: "Social housing was an important component of the planning strategy, permitted by the City's break-even financial policy on the land it owned and the grants of senior levels of government, provincial New Democrats and federal Liberals."[3]

Ley's thesis of a postindustrial society and the rise of a new form of liberal consciousness in the city was subsequently critiqued by leftist scholars on the basis of its orientation toward urban consumption and aesthetics rather than more fundamental processes and structures of change.[4] But his grasp of the dominant, socially liberal impulse of the 1970s, its ramifications for the urban spatial environment, and its subsequent demise with the rise of a "conservative" (neoliberal) hegemony in the following decade was important and far-sighted. In Vancouver, as elsewhere, the revanchist impulse of neoliberal urban policy was targeted precisely at this urban-professional cadre of reformers born in the freewheeling era of the 1960s. It was their ideological brew of social liberalism— comprising some economic redistribution, a modicum of social mixing, and the right of all residents to enjoy a pleasant, harmonious, and stable urban environment—that represented the enemy's front lines. The neoliberal advance guard targeted this hegemonic formation for "rolling back," well before the more aggressive state interventions of roll-out liberalism which followed.

Although Mike Harcourt, a social democrat, remained mayor of Vancouver between 1980 and 1986, a conservative provincial party was elected in 1980, with Bill Bennett as premier. At the federal level, another conservative party, the Progressive Conservatives headed by Prime Minister Brian Mulroney, took control in 1983. Following the rise of this configuration of neoliberal power, the attack on the former redistributive components of social liberalism was immediate. In Vancouver, the revanchist movement took shape in the deliberate "re-embourgoisement" of the city. This involved, first, massive state investment in land accumulation and real-estate development; second, the transformation of these sites into spectacles and festivals for the global elite; and third, the privatization, rezoning, and marketing of the property to offshore developers from Hong Kong.[5] The gentrification, disentitlement, and social dislocation caused by the mega–development projects which followed was immediate and intense.

The first dispossessions occurred with the loss of single-resident occupancy (SRO) hotels in downtown neighborhoods. This was accompanied by widespread rent increases and extremely low vacancy rates in Greater Vancouver in the late 1980s.[6] In November 1989, for example, the vacancy rate in Vancouver hit a low of 0.3 percent, with a particularly tight rental market in the West End and the downtown core.[7] Vacancies for studios and one-bedroom apartments were lower than for larger, more expensive apartments, affecting low-income single renters and single mothers with children particularly harshly.[8] In the outlying suburbs, vacancy rates were similarly dire. In some cities in the Fraser Valley, just 1 in every 1,000 apartments was vacant. The mayor of the Mission, which posted a zero vacancy rate in April 1990, spoke of the displacement of Vancouver residents to the outlying areas as the primary cause of the overcrowding: "The higher rents in the Greater Vancouver region have forced people who can't afford to pay them [to move] farther out." This statement was echoed by the mayor of Abbotsford, who noted that "a lot of people have moved from Vancouver to Abbotsford because our rents are cheaper."[9]

The average rent for a one-bedroom apartment in Greater Vancouver rose 9.4 percent in one year to C$548 a month in April 1990. (Average monthly rents in Abbotsford for a one-bedroom apartment were C$352 in September 1989).[10] Between 1988 and 1990, rental increases of 20 to 60 percent were common in downtown neighborhoods such as the West End.[11] Elderly tenants and women were particularly vulnerable, as seniors' incomes were insufficient to cope with the increases, and single women with children found it difficult to find alternative accommodations.

Furthermore, the heightened economic and social vulnerability of these two groups made them prey to harassment as they resisted the changes or sought new living quarters.[12]

The state role in facilitating this globally oriented gentrification was unparalleled. In terms of the urban imaginary, the city was "rescaled" as a key node in the global economy, a crucial urban gateway to the booming economies of the Pacific Rim, and a core city for Canada's ongoing development.[13] Using the rhetoric of globalization as both inevitable and desirable, local and provincial politicians reworked the image of Vancouver as a sleepy provincial town articulated most directly with its neighboring cities and hinterland into an image of a world city "naturally" connected most directly with Hong Kong and other key cities in Asia.

This Pacific Rim articulation, which also spurred a major capital inflow from eastern Canada and other regions of the world, swept the city into the vortex of global real estate and had an immediately deleterious effect on poor, working-class, and even middle-class residents of the city. The poorest segments of the population, residing in central city areas like the Downtown Eastside, were hit the hardest. They experienced the tail end of a domino chain of rising apartment prices and demolitions that thrust many of them from their single-residency hotel rooms onto the street. Many of the most disenfranchised in the city were former lumberjacks who had worked primarily in the False Creek sawmills before and after World War II. Ironically, these False Creek lands were among the first to experience global "regeneration" in the postindustrial, neoliberal city, causing the rapid gentrification of nearby property such as Chinatown, Gastown, Strathcona, and the Downtown Eastside, where these men lived.

With provincial control and offshore development, there are few local witnesses of power to the dislocations and social anomie caused by the regeneration and gentrification of the city. Offshore developers are not compelled by local issues of social reproduction to the same extent as local politicians and developers, who must negotiate with both their electoral constituency and the types of resistance detailed throughout this work. The growing tendency "for urban land to be treated as a pure financial asset and for managers in both the private and public sectors to pursue more intensively the rationality of rent and profit-maximizing" was clearly facilitated by the fundamental split between the logic of capital accumulation through *global* real-estate speculation, and *local* issues of urban consumption and social reproduction.[14] The neoliberal fantasy of the city as pure exchange value—spaceless and free-flowing, without inherent political meaning—was possible only through this form of deterritorialization.

In this classic scenario of contemporary urban transformation, the ongoing evocation of spacelessness emblematic of the neoliberal myth of the level playing field masked the state intervention that produced the spaces of spacelessness. A massive provincial role in land accumulation, development, spectacularization, privatization, and marketing was central in the transformation of Vancouver during this time—in addition to an active federal role in introducing new legislation such as tax incentives for business, and such as the Business Immigration Program, which enticed wealthy immigrants into investing in urban real estate. State intervention not only impacted the reorganization of city land, but also altered, in a mutually constitutive process, the relationships between capital and the state, and between the state and its citizens.

In a number of areas of urban life, the space of politics looked and felt quite different from the previous era. The differences experienced most directly at the urban scale related to the provision and consumption of social services, the protections (or lack thereof) afforded by local government, and the altered sense of entitlement, of hierarchy, and of a "natural" moral and spatial order. As I show, these experiences reflected the transition from a socially liberal perspective on the norms of urban life to a neoliberal orientation. It was (and remains) a transition deeply contested and unevenly implemented.

## Expo and the Restructuring of the Downtown

There has been growing agreement in recent years that Vancouver has the potential to become one of the few emerging "global cities." These are strategically situated urban areas endowed with favorable location, natural amenities, good climate and stage of development. They are capable of becoming interchange places with an economic base of management, communication and cultural exchange. . . . They are already becoming established in Vancouver.

   —*False Creek: A Preliminary Report*[15]

What is really at stake on this terrain is the heart of the city: the reconquest of the downtown for high-class users and high-rent uses. . . . In this view, revalorization is made possible by changing the city as a whole to a "higher" use, notably by converting it into a financial capital.

   —Sharon Zukin, *Loft Living*[16]

The image of Vancouver as a future "global city" was promoted in 1973 by the Planning Department's False Creek Review Panel. One year after the panel's recommendation, the Vancouver City Council upzoned ninety-five acres of False Creek land. This former industrial land, on the north shore of False Creek, comprised nearly one-sixth of the entire downtown area and was strategically located adjacent to the attractive False Creek inlet, and across from the successful south shore housing community built in the TEAM years of the 1970s.

   At the time of rezoning, the old industrial land was owned by Marathon Realty, a subsidiary of the Canadian Pacific Railroad.[17] The rezoning of the area from "Industrial" to "Comprehensive District-1" immediately increased the land's value by more than C$23 million, but Marathon refused to proceed with its planned housing development, and the land remained barren for several years.[18] From 1974 through 1979, the future use of this downtown land was endlessly discussed by city officials and Marathon agents. In 1980 the province finally purchased 176 acres of the False Creek real estate for C$30 million in cash and the exchange of more than C$30 million in downtown properties.[19]

   The saga of these old industrial lands reflects the restructuring of much of downtown Vancouver during the 1980s. Knowledge of the larger economic and political scenario is crucial for understanding the purchase, development, and marketing of this property. Owing to its reliance on resource exploitation, British Columbia was hit especially hard by the global recession of 1980. Unemployment grew from 6.8 percent in 1980 (88,000 people out of work) to 14.2 percent in 1985 (208,000 out of work) and was significantly higher than the national average. Income inequality in the province increased, and between 1981 and 1983, the number of households living below the national poverty line in British Columbia grew by 73 percent.[20]

The major political response to the crisis by the conservative Social Credit (Socred) provincial government was to introduce a "restraint" package, "a legislative 'mega-project' imposed in 1983 with the purported objective of addressing the world recession and stimulating the provincial economy through 'deregulation.'"[21] The proposed restraints focused largely on the excesses of labor rather than those of government, for while the working rights and benefits of labor were eviscerated, the provincial government engaged in a land-buying spree. In an attempt to stimulate the economy by diversifying away from resources and by initiating megaprojects, the provincial government during the early 1980s acquired massive quantities of land for redevelopment. These acquisitions included the 176 acres of former industrial lands abutting False Creek.

The False Creek land was purchased in 1980 for the development of B.C. Place, a provincial scheme that involved building a 60,000-seat stadium, office and retail facilities, thousands of housing units, and numerous trade and park pavilions. The B.C. Place project was connected with the incipient plans for an international exposition (then called Transpo) as a way to obtain both federal funding and international recognition for the city.[22] Premier Bill Bennett made explicit the planned festival's connection to overall future growth and massive government-led development in the city, saying of the initial project that "the trigger for [B.C. Place] development will be Transpo '86."[23]

And sure enough, the False Creek land purchase and B.C. Place redevelopment project stimulated a number of major development schemes throughout the city. The Expo '86 theme, "World in Motion—World in Touch: Human Aspirations and Achievements in Transportation and Communication," emphasized local and global linkages, particularly those facilitated by new transportation technologies (hence the festival's original name). Drawing from this useful theme, the Bennett government strategically argued that an exposition featuring transport would appear ridiculous in a city without a modern transit system. The federal government and several million taxpayers coughed up C$1.2 billion for the new automated light rapid transit system, ALRT (Skytrain), thus immensely improving circulation and integration within the city.[24]

The festival itself, Expo '86, occurred with great fanfare and made much of the "pride of place" theme common to international expositions. Besides the transportation and communication motifs, the exposition theme that was frequently stressed was Vancouver's coming of age as "a centre strategically placed on the Pacific Rim, the new development axis."[25] The international character of Expo '86 was celebrated in a reception for international students and diplomats in October 1985, where guests were treated to strawberries and wine and shown Canada's burgeoning capacities in high technology. This reception occurred just two days after a meeting at the Carnegie Community Centre, where residents of gentrifying downtown neighborhoods voiced anger and dismay at rent increases and eviction notices to Expo's vice president for communications, George Madden.[26] Pred wrote of this type of disjuncture in Sweden:

> The educational exposure of international exhibition visitors to the objective possibilities of new forms of consumption was one with their exposure to representations that glorified domestic industrial achievements. . . . At the same time, such exposure was one with an exposure to ideological articulations that linked industrial technology with continuous progress and a belief in an ever better future, to ideological articulations that silently denied the widespread existence of domestic tensions and poverty, and thereby legitimated the present.[27]

The festival's glorification of domestic achievements and denial of domestic tensions was initially quite successful. Nine days after the closing of Expo '86, a new Socred government, under the premiership of William Vander Zalm, won the provincial election. Vander Zalm immediately initiated a sweeping land-privatization campaign that necessitated the sale of vast quantities of undeveloped and enormously lucrative land in British Columbia, including the Expo site itself. Under the new privatization strategy of the Vander Zalm government, these lands, acquired at great cost under the Bennett government, were scheduled for sale back to the private sector. In March 1987, under the direction of the province, the B.C. Place Corporation and the B.C. Development Corporation (two companies that had been responsible for the development of the Expo site and other provincial lands) were merged into a single entity: the B.C. Enterprise Corporation (BCEC). The raison d'être of BCEC was to "divest itself of the Expo lands, and of holdings in fifteen other B.C. communities, including property in Whistler and Victoria."[28] In May 1987, Grace McCarthy, the new minister of economic development, stopped the proposed provincial development on the eastern end of the Expo lands, saying that it was uneconomical. Soon after, the BCEC staff began to interview developers interested in acquiring the site wholesale. On July 4, 1987, BCEC president Kevin Murphy met with Hong Kong billionaire Li Ka-shing to discuss the possible purchase of the Expo lands. After a few meetings between Li and McCarthy, and after much unsuccessful intervention by Vander Zalm on behalf of his friend the local developer Peter Toigo, the BCEC board approved the sale of the Expo lands to Li Ka-shing in April 1988.[29]

The central role of the provincial government in the restructuring of Vancouver's urban downtown thus continued with the sale of a large portion of it. After amassing the False Creek land and creating the spectacle that helped commodify it, the province sold the property to the targeted buyer, a major capitalist from Hong Kong. Thus the government's strategic manipulation of the land throughout the 1980s set the stage for its grand dénouement. By displaying its urban wares on the international stage and promoting the themes of progress and integration in the Expo extravaganza, provincial leaders convinced the federal government that Vancouver was a new major player on the world scene, eminently deserving of vast quantities of federal support. With federal money in hand, massive infrastructural projects were begun which further enhanced the city's spatial integration. The improvement of transportation and communications technology made the city an increasingly attractive place in which to invest. By the late 1980s, Vancouver real estate became for a time the hottest property in the world.

In this context, the sale of the Expo lands to Li Ka-shing can be seen as a deliberate move to increase the city's global integration by attracting capital from Pacific Rim investors. By nixing government development on the land, and by canceling earlier plans to sell the land in smaller parcels, Minister McCarthy ensured that only the largest development corporations were able to bid for the site. Two bids were tendered—Li's winning bid and one by the local developer Jack Poole, of the Vancouver Land Corporation (VLC). The sale price to Li was C$320 million, with a C$50 million deposit and C$10 million a year for five years beginning in 1995; then C$20 million, C$40 million, C$60 million, and C$100 million over the next four years. The bid, considered low by many outsiders, was accepted with alacrity by provincial leaders. Whether or not the province was receiving top dollar for the lands was secondary to the expected business opportunities generated by the purchase of the site by Li Ka-shing. The "world city" future planned for Vancouver was candidly acknowledged by the leading politicians involved in the deal. Minister McCarthy said of the dinner she shared with Li *before* the sale: "There was talk about business opportunities in B.C. and about developing Vancouver as an international finance center."[30]

Selling the downtown land was an attempt to sell Vancouver as a viable investment center to wealthy capitalists from Hong Kong and other Pacific Rim regions. The strategy stemmed from an awareness that Li's reputation for having the Midas touch was so high in Hong Kong and elsewhere that fellow investors often followed his early lead in development projects around the world. The accuracy of this assessment was confirmed by Sam Yip, a lawyer with Stikeman/Elliot, the law firm engaged with Li's global operations. Yip noted that people in Hong Kong invested heavily in Vancouver real estate only *after* Li's purchase. He went on to add that at the time of the original bid, no other credible international group (such as a Japanese consortium) had offered to buy the site. Vancouver, he said, was "not on the international scene yet," and it was necessary for "some big name to take the risk." According to Yip, the subsequent rise in the value of Vancouver real estate, including the Expo property itself, was a direct function of Li's purchase: "Now Vancouver is on the international map, so of course the land is worth more. The Expo site transaction was reported worldwide. This was the first time it was so widely covered. Why? Because of Li's name and the size of the investment. The deal put Vancouver on the map internationally."[31]

During the preliminary stage of the land agreement with Li, Minister McCarthy wrote a letter to Vancouver city councilors outlining a bonus structure for the city if it allowed higher densities on the Expo site. Li's original bid of C$320 million had been tendered and accepted by all parties regardless of the eventual zoning mandated by the city. The bid, however, included a provision whereby the province could collect up to C$500 million (C$180 million *more*) for the deal if Concord Pacific was allowed higher building densities on the site. McCarthy pointed out in her letter to the Vancouver City Council that although the province would take in the extra money, the city would reap a significant financial payoff from this upzoning as "a generous gesture of good will" from provincial leaders.[32] Further provincial shenanigans were exposed less than a year later, when the publication of the original contract from the Expo sale showed that the B.C. government had been willing to amend the Land Title Act so that waterways proposed on the site by Concord Pacific would not remain publicly owned.[33] Presumably this perquisite had been offered to Li to allay fears that public ownership might interfere with his company's proposed developments on the site.

The provincial attempt to bribe the city for higher density and amend legislation that protected the city's control over its waterways manifests the high value that the provincial government placed on consummating the sale with Li. In 1989, as more and more Expo land deals were uncovered by irate Vancouver journalists, Minister McCarthy admitted that BCEC had been under great pressure from the provincial cabinet to sell the site quickly in 1988.[34] The sacrifice of Vancouver's livability and of the ability of its residents to control the process of change was considered an unfortunate but necessary side effect of the scheme to attract Asian capital. When it became generally known that the price of the Expo land had quadrupled by 1990, and that Li Ka-shing sold a tiny 4.2-hectare plot of land to a Singapore developer in February for C$40 million, city residents and local politicians grew increasingly furious.[35] When it became further known that the federal and provincial governments were forced to *buy back* a portion of the land (at C$28–$30 a buildable square foot) for social housing, opposition to the Expo land deals became virtually citywide.[36] One journalist wrote in an op-ed piece entitled "Let's Buy the Whole Mess Back":

> "Invite the World" was the Expo theme. What better way to carry that on than by attracting really big, high-profile money. Would not many other smaller investors follow Mr. Li to Vancouver? Indeed they might. Indeed they have. Now, some of us in Vancouver think that the world is arriving here a little faster than is good for us. We are growing in

Vancouver at over 3 percent per year. That doubles in 24 years. That is too fast, in terms of housing costs, infrastructure, racial frictions, and other features of rapid change.[37]

The Expo land was amassed, privatized, and sold by provincial officials who consulted little, if at all, with Vancouver city and community leaders and residents. As the real-estate scandals that involved Premier Vander Zalm were uncovered, and as real-estate and development links that involved Minister McCarthy and her husband began to emerge, city residents' perceptions that they had been sold out by the Socred government crystallized. Anger centered on the role of the state in the deliberate commodification and sale of the city to the highest bidder. Many people connected individual politicians' Hong Kong liaisons and trips to Asia with the sellout of the city. A furious Vancouverite wrote to the editor of the *Vancouver Sun* on February 20, 1989:

* Blame the Asian Invasion on Premier Who Invited It. *

Who has been inviting investment from Hong Kong actively for three years? Who has been going on "fact-finding" economic junkets on your tax dollars in Hong Kong to invite business to bring their dollars? Who sold the Expo lands at below market value? The provincial government, that's who. In league with the business community, they have been begging, pleading and greasing the path to B.C.

The anger evident in this letter demonstrates some of the perils of neoliberal policy for local and provincial politicians. It also reflects the imminent outbreak of overt racism, as particular areas of the world (such as Hong Kong) and the perceived agents of capitalist dynamics of change (such as the Hong Kong Chinese) were increasingly held responsible for the negative effects of urban transformation. When "crises" in social meanings, as well as in the material bases of social reproduction, flare up in this manner, it threatens the carefully cultivated neoliberal image of spacelessness and of the harmonious planetary embrace of global capitalism. This is especially the case with respect to efforts by state officials to attract foreign investment and foreign *investors* to the metropolis.

In moments such as these, there is a strong effort by state actors to enhance cultural articulations and produce the types of institutional connections that galvanize hegemonic formation. This effort includes the formation of transnational cultural ties.

In Vancouver's case, state actors at the provincial and municipal levels of government attempted to create a smoother neoliberal surface in the city (i.e., one not disfigured by racial antagonism and other kinds of blockages) by establishing cultural festivals in both Hong Kong and Vancouver in the early 1990s.

## THE CULTURE OF INTEGRATION

We are pleased to see the increasingly close links between Canada and Hong Kong. Not only are we each other's major trading partners, but also good partners with growing ties in the fields of culture, recreation, and education. . . . Through this two-way exchange of festivals, we hope to reinforce and improve both the business links and peoples' links between the Hong Kong and Canadian communities.

 —John Chan[38]

Festival '91, the exhibition of friendship and international cooperation, was launched amid great hoopla during Prime Minister Brian Mulroney's visit to Hong Kong in the spring of 1991. The festival, whose theme was "Hong Kong: Friends Yesterday, Today, and Tomorrow," was intended to help "foster good relations" between Hong Kong and Canada.[39] Following the Canadian festival in Hong Kong, a Hong Kong festival in Canada was scheduled for 1992. Each offered seminars, concerts, dance, food, films, and cultural exhibitions.

Festival '91 was the brainchild of Hong Kong governor David Wilson. On a trip to Canada in 1990, he spoke with Prime Minister Mulroney about the need for a promotional event that would encourage the growing links between the two regions. These links, including the movement of capital and people from Hong Kong to Vancouver in the late 1980s, had already produced racial friction between the two cities, and the perception of an increasingly negative business climate in Vancouver. What was needed to counteract these perceptions, Wilson felt, was an event that promoted the image of a *positive* exchange between the two locales, an event such as a festival, a spectacle of ongoing and harmonious cultural relations.

Festival '91 was heavily advertised in Hong Kong, and small exhibitions and displays were set up throughout the city during June. Each display emphasized global interconnectedness and the

importance of establishing and maintaining economic and cultural ties between Canada and Hong Kong. (For example, a display by ALCAN at the Tai Koo Shing Mall demonstrated the
importance and interconnectedness of aluminum recycling for
the people of Hong Kong and Canada.) Canadian food delicacies
such as smoked salmon were displayed in a number of shops, and
recipes for unusual Canadian dishes appeared in magazines and
storefront windows. The festival's main events highlighted cultural and educational programs; the scheduled entertainment
included a piano recital, a cello concert, and a dance performance
by Canadian artists.

The festival, which cost approximately C$4.2 million, was
funded by the Canadian government, and by several local Hong
Kong patrons and sponsors.[40] These sponsors were uniformly
implicated in massive investment schemes that involved the
articulation of Hong Kong and Canadian capital. The Platinum
Maple Leaf Patrons, who donated C$100,000, for example,
included some of the major business interests operating between
Hong Kong and Vancouver in the late 1980s: Li Ka-shing, Stanley Ho, Semi-Tech (Global), Ltd., Canadian Airlines International,
Ltd., and Manulife Financial. Other patrons and sponsors included
major banks, real-estate and development corporations, and investment companies with similarly transparent interests in furthering the Hong Kong–Vancouver connection.[41] For all these
patrons and sponsors, smooth economic relations between the
two regions were clearly linked with the extension of good cultural relations.

The emphasis on cultural relations served a dual purpose. By
selling the so-called peoples' links, the festival helped downplay
the massive amounts of capital being transferred from one region
to the other. The ramifications of this rapid capital inflow for Vancouver's built environment were immense, and were largely
responsible for the growing racial friction in the city toward the
end of the 1980s. Cultural exchange also helped disguise the mercenary nature of the relationship between the cities. Further, the
promotion of cultural ties and the insistence on harmonious international relations were important for the connections between
businesspeople. Hong Kong capitalists interested in investment
opportunities in cities such as Vancouver were reassured that

racism would not block capital circulation or affect them personally. And Vancouver citizens, angered at a perceived sellout of
passports for dollars with the new Business Immigration Program,
were assuaged by the emphasis on a salutary cultural exchange.

The festival was presented as part of an "international village,"
with a strong rhetoric of globalization as both inevitable and
desirable, and cosmopolitanism as achievable only for those able
to establish and maintain international linkages. In the official
festival program published in *Canada–Hong Kong Business* magazine in June 1991, representations in the advertisements reflected
this new neoliberal mandate. The advertisements stressed the
growing interconnections in the world and the imperative to be
"hooked up," "networked," "linked," and otherwise imbricated
in the international networks of the global economy. Hutchison
INET, a telecommunications subsidiary of Hutchison Whampoa,
for example, showed a man casting a net into the sea. The text
read: "How well you succeed offshore depends on how far you
extend your network." Manulife Financial, an international insurance firm, juxtaposed the Hong Kong waterfront and the Vancouver waterfront, bays and skylines running together in the middle of the page. The text read: "We make the connection between
Canada and Hong Kong by serving business and family clients in
both countries."

In Festival '91, state-led articulations of the friendly and inti-
mate "international *village*" posited a harmonious cultural part-
nership at the same time they assured businesspeople of a sim-
patico economic climate. The government's welcome extended
to the country as a whole, but the economic restructuring presented so benignly in the festival brochures masked a desperate
and highly competitive scramble by individual provinces and
cities for the increasingly mobile Hong Kong capital. Flexible
accumulation opened up new paths and new opportunities for formerly small-town cities like Vancouver to lure international
investment; at the same time, it forced them to compete with
other cities and regions for the goods. Harvey writes of the instability of this position: "Small towns that have managed to lure
in new activities have often improved their position remarkably.
But the chill winds of competition blow hard here too. It proves

hard to hang on to activities even recently acquired. As many cities lose as gain by this."[42]

In addition to Festival '91, there were a number of less publicized economic and cultural exchanges between Hong Kong and Canada during that year. In a festival labeled, appropriately, Money '91, investment was the unabashed focus. Yet even in this milieu, where business opportunities and economic ties were foregrounded, the superior cultural attractions of Vancouver remained central. A film shown to the assembled group of eighty or ninety well-dressed businesspeople enumerated the benefits of investing in British Columbia: the Free Trade Agreement (FTA) with the United States, low energy and labor costs comparable with those of Oregon and Washington, the second-largest port in North America, and various advantages associated with a high-quality lifestyle, including good postsecondary education, low crime, and cleanliness. After the film, Dickson Hall, the government representative for British Columbia, answered questions from the floor.

In an interview in Hong Kong in April 1991, Hall reiterated to me the explicit intent of the province: to attract business investors to British Columbia. He noted that Vancouver was virtually synonymous with British Columbia, as 95 percent of trade and immigrants to the province moved to this central city of the lower mainland. Hall's office in Hong Kong (the B.C. Foreign Trade Office) contained materials on investing in and emigrating to British Columbia, including materials disseminated by banks, investment funds, manufacturing companies, and law firms. The focus on attracting *wealthy* immigrants was manifest in the preponderance of information related to money and investment. Government pamphlets emphasized, in particular, the new Business Immigration Program. In one brochure, "Your Future Is in British Columbia Canada," Premier Vander Zalm and Minister of Economic Development McCarthy addressed messages of welcome to prospective investors. Much of the rest of the pamphlet focused on a description of economic regions and investment opportunities in the province and on the innumerable reasons for choosing British Columbia over other Canadian provinces or the United States. As in the seminar, issues of lifestyle, heritage, climate, and worldwide connections were heralded as primary reasons to invest in British Columbia.

Another pamphlet that sought to attract business immigrants was entitled "Welcome to British Columbia, Canada: The Business Immigration Program." On page 2 of this brochure, the minister of international business and immigration, Elwood Veitch, wrote to prospective investors:

> British Columbia is truly Canada's "Opportunity Province." With the Canada-U.S. Free Trade Agreement in place, our increasing role in the Pacific Rim, our strong traditional trading links with Europe, and our growing ties with many other nations, our province has become a major player on the international scene. The Ministry of International Business and Immigration looks forward to working with business immigrants who are interested in making British Columbia a home for themselves and their investments.

The minister's emphasis on established global connections was calculated to convince investors that Vancouver was spatially well integrated into the advanced system of international capitalist networking. Capitalists who wished to invest in British Columbia could rest assured that their financial portfolios could be plugged in with relative ease. The message of economic preparedness and receptivity was delivered repeatedly, with advertising in Festival '91, Money '91, the Asia Pacific Foundation symposium, the British Columbia office, *Canada–Hong Kong Business* magazine, and many other forums. As Prime Minister Mulroney remarked on May 24, 1991, to the Canadian Chamber of Commerce: "Billions of dollars of investment from Hong Kong are contributing to the dynamism of both our economy and yours. . . . Canada's past has been largely an Atlantic past. Canada's future will be increasingly a Pacific future. We want Hong Kong to play a major role in that future."[43]

The effort to sell Vancouver as a world city did not originate with Festival '91, although it began to take shape with an exhibition of a similar nature. The integration and marketing of the city was part of a broad, long-range strategy that first gained momentum in the early 1980s and grew rapidly with Expo '86. Seven months before Expo took place, six members of a Canadian delegation flew to Hong Kong to promote the festival there. Claude Richmond, the minister of tourism for British Columbia, stated that Expo's promotion was designed to "build interest in Canada's provinces as places for business and investment."[44]

The Canadian delegation arrived in Hong Kong in 1985, just one year after the signing of the Sino-British Joint Declaration, and one year before Vancouver's international exposition. The timing of the visit was not accidental. As Hong Kong's future was negotiated by colonial politicians and their neighbor to the north, Hong Kong businessmen considered the options available for their capital investments. Many regions around the world attempted to attract the Hong Kong capital, but British Columbia laid siege in an especially stubborn and effective manner. Building on preexisting connections and introducing immigration programs designed to entice business investors, the province began an all-out campaign to sell Vancouver as a world-class city in which to do business and reside.

One of the first politicians to court Asian capital was Vancouver mayor Mike Harcourt. Harcourt made several trips to Asia to promote the city and to meet with potential investors while he served as the mayor between 1980 and 1986. When he was elected premier of British Columbia in October 1991, he returned to Hong Kong within three weeks of the election to reassure wealthy international investors that the (left-leaning) NDP remained committed to furthering the Pacific Rim connection. He said in response to a question about British Columbia's burgeoning trade with Pacific Rim countries: "Yes . . . and it will grow too with South America and Mexico. That is our role, British Columbia's role, in the medium and long term, to be Canada's front door on the Asia Pacific, to act as a catalyst and as a meeting place. This is where the action is."[45]

## Neoliberal Strategies of Global Integration

In addition to the provincially mandated urban policies of land privatization and targeted marketing in Hong Kong, a number of federal policies affected Asian investment in Vancouver. The Business Immigration Program, for example, was created in 1978, redesigned in 1984, and extended in 1986. It was expressly designed to facilitate the immigration of capitalist entrepreneurs who could "make a positive contribution to the country's economic development by applying their risk capital and know-how to Canadian business ventures which create jobs for Canadians."[46] The program was targeted at Hong Kong immigrants who might be considering a move out of the colony in advance of the transition to Chinese control in 1997. As one University of Toronto economics professor baldly put it: "The wealth creators are in another place, and we must import them. There is no alternative."[47]

In 1986, entry into the program was divided into two categories, "investor" and "entrepreneur." As a condition of entry, investors were required to have a minimum personal net worth of C$500,000, obtained by their own efforts, and to invest at least half that amount in a business or privately administered investment fund in a Canadian province for at least five years.[48] For acceptance into British Columbia, the investor had to commit C$350,000 for a three-year period to a project approved by the Canadian government.[49] The object of the investor category, from the government's perspective, was "to attract qualified business persons to Canada on the basis of their willingness to invest their capital in Canadian business ventures which create jobs and contribute to business expansion."[50] The entrepreneur was required to have proven management skills and a business background, and to actively manage the business in Canada once it was established. Furthermore, the Canadian business had to contribute significantly to both the Canadian and the provincial economies, and employ one or more Canadian residents other than the entrepreneur.[51]

Applicants accepted under one of these business categories were given queue priority and extra points in the processing of their applications by Canadian immigration officials.[52] The huge net increase in immigration from Hong Kong to British Columbia in the late 1980s reflected the success of the federal government's strategy. From 1984 through 1991, Hong Kong led as the primary source country under the program, jumping from 338 landed business immigrants in Canada in 1984 to 6,787 by 1990.[53] (The total landings from Hong Kong in 1990 were 29,266, up nearly 10,000 from the year before.) By 1991, the number of visas issued to Hong Kong residents in the business class grew to 8,159.[54]

The number of business immigrants who chose British Columbia as their final destination in 1989 and the first six months of

1990 was significantly higher than for any other province in Canada. In 1989, of the 1,121 entrepreneurs who departed Hong Kong for Canada, 384 (36 percent) landed in British Columbia. In the same year, of 406 investors, 177 (44 percent) chose British Columbia.[55] Figures from *BC STATS* show even higher numbers.[56] As with the general immigrant population from Hong Kong, the vast majority of business immigrants to British Columbia opted to settle and invest in Vancouver. Between 1988 and 1990, the number of people emigrating from Hong Kong to Vancouver continued to rise. In 1988, 4,965 of 5,188 landings in British Columbia were in Vancouver (95 percent); in 1990, 7,471 of 7,660 (97.5 percent).[57] The number of people arriving in Vancouver from Hong Kong continued to increase in the early 1990s, with the largest group, 15,663, entering in 1994, and an average in-migration of more than 10,000 between 1990 and 1996.[58]

The rules that regulated the Business Immigration Program were instrumental in creating a highly transnational or "flexible" group of migrants. For example, the stipulation for those entering under the entrepreneur category included the proviso that the migrants create or maintain a business employing Canadians for a period of five years. To obtain citizenship in Canada, they were further required to live in the country for a minimum of six months out of every year over a three-year period. However, most of the migrants' personal resources and business networks, and often their fixed capital investments, remained in Asia. In addition, aside from investment in real estate, business opportunities were greater in the booming economies of the greater China region than in British Columbia in the late 1980s and early 1990s.

The rules of the program thus required entrepreneurs and investors to commit time and resources to doing business in Canada, yet to be successful in this pursuit, many immigrants needed to maintain their close ties and expand their opportunities in Hong Kong, Taiwan, or China.[59] As a result, a great number of migrants were forced to remain essentially mobile and operate as global economic subjects living part-time in Vancouver, part-time in Hong Kong. They spent so much time flying back and forth they won the nickname *tai hong yan*, or "astronaut," which functions as a pun and carries the double meaning "empty wife" in Cantonese, a reference to the limited amount of time these businessmen were able to spend with their families.[60]

The Business Immigration Program was part of a much broader, federal-scale, neoliberal agenda in the 1980s that emphasized the liberalization of finance and the establishment of capital networks around the globe, particularly in the booming economic region of the Asia Pacific. This agenda crystallized in the ratification of two free-trade agreements, the FTA (Canadian-U.S. Free Trade Agreement) in 1989, and NAFTA (North American Free Trade Agreement) in 1993. Key aspects of this neoliberal agenda included the decentralization and attrition of federal governance, the accordance of a greater degree of power and control to provincial authorities, the deregulation of banking and other institutions, the privatization of land and industry, the reduction of friction for the free circulation of commodities, and the provision of various tax and other incentives attractive to business. In the 1980s these institutional transformations manifested a new direction for Canadian society. Moreover, the changes promoted during this time in trade, banking, and the bureaucratic organization of government occurred alongside a rhetoric of national deficit and decline firmly linked with the "excesses" of welfare-state provisioning under Trudeau's liberal government of the 1960s and 1970s.[61]

The reduction of friction to the free circulation of capital and commodities was only one feature of the rise of neoliberalism in Canada. An important corollary to the liberalization of capital was the attrition of the social service provisions associated with the welfare state. The most obvious attack on federal provisioning and the preexisting organization of government was evident in the attempted changes to the Canadian charter and constitution. These changes, while usually discussed in terms of cultural and regional disputes in Canadian politics, especially vis-à-vis the constitutional accords and Québec, were primarily concerned with the manner in which distribution occurs in a federalist system.[62] The various disputes and attempted constitutional changes between the 1960s and the late 1980s—including the debates around cultural dualism, regionalism, the Meech Lake Accord, and the Charlottetown Accord—did not represent merely struggles of identity, decentralization, and cultural or regional independence, but were more broadly related to reordering social and federal-provincial institutional relationships and reducing government to make way for the private sector.[63]

   The conservative attack on the provisions of the Canadian welfare state was evident in a number of areas. These included a regressive (and highly unpopular) goods and services tax, more stringent eligibility rules for unemployment insurance, a deindexation of family allowances, freezes on federal spending in areas such as education and health care, and a cap on Canada Assistance Plan payments, which operated as a strong disincentive for provinces to increase social service spending.[64] This cap was followed by a 1991 budget freeze in federal government payments on Established Program Funding (EPF) for social services such as Medicare and postsecondary education.

   At the same time federal funding for services was frozen or capped, the very ability of the government to supply funds was eroded. During the 1990s the power of the federal government to tax citizens for social service programs devolved steadily to the provinces. This downscaling of the federal role in social services was particularly important for the formulation of politics in British Columbia. The rise of the Progressive Conservatives in federal politics was paralleled by the growth of a staunchly conservative provincial force, the Socred Party. The Socreds embraced the move toward privatization and deregulation. Following premier Bennett's election in 1983, the party immediately reduced the number of public service employees on the payroll and cut spending for social services. This was followed by a strong rhetoric of the necessity to reprivatize social services to promote maximum efficiency.[65]

   The strategies of land accumulation and then privatization in Vancouver took place within this broader economic and political climate. So did the numerous fairs and other extravaganzas, which presented Vancouver as a spectacular city ready to be swept up into the networks of global real estate. In this manner, the city was quickly shaped in ways amenable to the economic and spatial restructuring that was occurring on national and global scales. Although this is clearly not a new phenomenon for cities, as Smith has argued, "what *is* new today is the degree to which this restructuring of space is an immediate and systematic component of a larger economic and social restructuring of advanced capitalist economies."[66]

   The impact of a large influx of capital and immigrants from Hong Kong during this time was particularly noticeable in Vancouver because of the city's small size and provinciality. Figures from 1989 show an approximate capital flow of C$3.5 billion from Hong Kong to Canada, of which C$2.21 billion or 63 percent was transferred by the business migration component.[67] However, these figures are quite conservative, considering that most applicants underdeclare their actual resources by a significant margin to evade taxation.[68] Bankers and immigration consultants I interviewed in Hong Kong in 1991 put the overall numbers as high as five or six billion Canadian dollars transferred from Hong Kong to Canada annually in the late 1980s and early 1990s. (A few believed the figure to be much higher.) Of that amount, over one-third was destined for British Columbia.[69]

Immediately following the announcement of the immigrant-investor category in January 1986, a host of consulting firms sprang up to "educate" the potential Hong Kong emigrants about visa requirements and Canadian investment opportunities. At the same time, innumerable investment funds were formed, such as the Canadian Maple Leaf Fund, the Beacon Group of Vancouver, the Merbanco Pacific Group, all aiming to take advantage of the new immigration law. The Maple Leaf Fund, initiated by Stephen Funk, chair and president of First Generation Resources, Ltd., an investment-banking company, was the first and one of the most successful of the private investment funds to combine immigration concerns with a trust-fund offering. Funk said of his fund: "We're here to help investors take advantage of Canada's immigration laws. But we're going to manage this fund as a growth fund, and that's the most difficult kind of fund to manage."[70]

The banks also initiated a number of investment funds targeted at potential Canadian immigrants. In early 1986, the Canadian Imperial Bank of Commerce (CIBC), First General, and a number of others opened investment funds in Asia that would qualify investors to apply for residence in Canada. The early CIBC funds, which were fairly standard, were federally sponsored venture-capital funds aimed at boosting investment in the provinces of British Columbia, Ontario, Québec, and Alberta.[71]

All the banks attempted to attract investors to these funds by offering a wide range of services connected with general immigration concerns. The Bank of Montreal, for example, provided Chinese-speaking customer-service people in Canada who were "especially trained to respond to the needs of offshore and immigrating investors."[72] The Hong Kong branch offered to open accounts in Canada, accept funds for deposit in Canada, and assist with mortgage financing for Canadian residential real estate, along with brokerage, emigration, and emigration-trust services. The bank's emigration-trust services were set up through Harris Trust and Savings Bank, a wholly owned U.S. subsidiary of the Bank of Montreal. As a bank executive told me in a May 1991 interview in Hong Kong, clients could use trusts in the United States to avoid Canadian taxes for a period of five years after emigration, a common strategy for wealthy overseas executives as well.[73] Other international services of the bank included assistance in making major capital deposits in its Singapore branch to avoid both Canadian taxes and the possibility of nationalization in Hong Kong after 1997.

In addition to attracting new investors, the personal services offered by the bank to assist people in moving from Hong Kong to Canada were structured largely to retain the bank's established clients. The same senior bank official told me quite emphatically that many of the personal services offered by the bank, including entertaining clients during their trips to Canada, were based on the imperative "not to lose clients." Several executives in other banks also mentioned the need to keep a good relationship with their clients by helping them to establish a new life in Canada. A senior executive at Citibank told me in an interview in Hong Kong in May 1991 that the number-one priority of the bank was to give good service to their clients so that they would transfer to one of the bank's overseas branches rather than shift to another bank.

Customers in Citibank's Hong Kong emigration program each deposited between U.S.$1 million and $10 million, with a minimum investment of U.S.$1 million per account. Many of the bank's clients were fabulously wealthy; the official I spoke with believed that most of the clients who emigrated to Canada transferred only 10 percent of their total funds at the time of their move. He believed that the other 90 percent would eventually wind up in Canada, but "a lot depends on what happens in Hong Kong in the meantime." Most of his clients' money was not located in Hong Kong but diversified in investments and portfolios around the world. The imperative to remain involved in the movement of these enormous amounts of capital (and people) ensured that bank executives offered every possible service, and that the banks expanded in size and geographical networking to accommodate their clients' new needs.

The largest bank to facilitate the increasing movement of people and capital from Hong Kong to Canada was the Hong Kong and Shanghai Banking Corporation (Hongkong Bank). David Bond, the vice president of the Hong Kong Bank of Canada, its Canadian subsidiary, said of the business investors who were moving to Vancouver: "If I was the czar of immigration, I'd send a fleet of Boeing 747s to Hong Kong to pick them up. This is a unique chance to engage in a transfer of human and financial capital that is unprecedented anywhere in the world."[74] Following this line of reasoning, the Hong Kong Bank of Canada expanded at a tremendous rate in the 1980s, aided by the reputation of the parent company and its long-term relationships with a well-heeled clientele.[75]

All the major international banks that operated in both Canada and Hong Kong emphasized their global connections and services to attract and retain customers. The vice president of the Royal Bank of Canada claimed that, "with operations in more than 30 countries, we can provide worldwide linkages for the investment and personal interests of our private banking clients."[76] Like the Bank of Montreal, the Royal Bank of Canada offered "premier V.I.P services" for its Hong Kong clients, including "a comprehensive package of services to meet all your emigration financial requirements."[77] Help in real-estate ventures, both personal and commercial, was particularly emphasized. Soon after the new immigration regulations of 1986 were in place, for example, the Royal Bank executive in charge of private banking operations established an investor-immigrant service to help wealthy clients purchase real estate in Canada and to structure five-year offshore tax programs. The first phase of the service package provided for fund transfers and the opening of new accounts in the bank's main branches in Canada; in the second phase, the package provided for mortgage financing of Canadian properties.[78]

Many bank officials estimated that a majority of the funds transferred from Hong Kong to Canada were invested in property. Some trust companies, which also became involved in the immigrant-investor scenario, gave direct assistance on both immigration and property investment through subsidiary or sister companies. Royal Trust Asia, Ltd., for example, acted on real-estate matters through its sister company Royal LePage, the largest real-estate broker in North America. By combining departments or subsidiary firms, companies were able to provide services tailored for the wealthy Hong Kong emigrants, and to get a jump on the property and development markets at the same time.[79]

Broad trends in the global economy that had a particularly strong impact on real-estate investment occurred in the financial arena. The rise of nonlocal sources of development financing, accompanied by a widespread deregulation of financial markets, facilitated investment in new kinds of commercial and residential developments not just in Vancouver, but worldwide.[80] The liberalization of international finance during this time had major implications for investment in the Vancouver built environment. With the rapid and extensive restructuring of the financial system, the state's control over money supply, allocation, and value declined, and money became increasingly mobile and unconstrained. Historically, the regulation of Canadian banks took shape under a system known as the Four Pillars, which involved the separation of financial activities into the four categories of insurance, banking, securities, and trusts. Liberalization of the industry in the 1980s included the removal of the numerous state-based restrictions governing the separation of these financial activities, as well as the termination of ownership restrictions on financial institutions.[81] Thus, at the same time federal immigration laws greatly eased the conditions of entry and citizenship for wealthy business migrants, federal deregulation of the banking industry eased the movement of capital across both international and sector borders.

Following the deregulation and liberalization of the financial system, banks began to dominate all fields of financial activity except insurance, and quickly began to consolidate partnerships both domestically and offshore. Investments in Vancouver property in the 1980s were greatly affected by financial deregulation, owing to the new forms of finance capital available for property acquisition. Insurance companies and pension-fund investments in real estate, for example, grew markedly as deregulation helped remove institutional barriers that had separated investments in residential development from those in nonresidential development and in stocks and bonds.[82] At the same time, new mechanisms of securitization that linked real estate to broader capital markets made many of these property investments increasingly attractive. With declining national control over both physical and financial borders, and with an increasing emphasis on nonbank sources such as international equity and bond markets, global investors were able and willing to speculate in property development on a hitherto unprecedented scale.[83]

The rise of nonlocal sources of development financing and the concurrent globalization of property markets played a major role in the transformation of cities worldwide. For Vancouver, the social and economic effects of large-scale investments in real estate were shocking, even for jaded social activists and urban historians. Donald Gutstein, a journalist, author, and long-time Vancouver resident who has chronicled the impact of successive waves of development in the city since the 1960s, said in our November 1990 interview in Vancouver:

> What has really made an impact on me is the globalization of real estate. I've actually seen what it could mean here. Certain parts of Vancouver are being drawn into this global real estate market where investors could be anywhere. They've decided that this part of Vancouver and that part of Toronto and that part of Brussels and so on are worthy to be considered part of an investment portfolio. So prices in those areas are totally taken out of local hands. They no longer respond to anything that's happening in Vancouver. The nation-state is obsolete with the globalization of the economy. You have this wide network of cities like New York and Paris and Hong Kong and Tokyo which drive the economy and between which capital and information flows, and Vancouver is being drawn into that, sort of as a subsidiary.

Despite government attempts to channel capital into productive sectors, the majority of Business Immigrant Program funds, particularly in the early years, went into property investment. Rather than creating businesses and employment for Canadians, much of the capital invested in early funds subsidized developers in preplanned business ventures. One journalist from Vancouver noted: "In practice, a lot of the C$1.1 billion thus far has served as replacement financing, subsidising developers whose projects would have been built anyway. A hotel builder, for example, may create an immigration fund and take in partners at C$250,000 each, paying them eight percent interest for money that would cost 12 percent at the bank. No new venture is created, and the developer pockets the four percent spread."[84] Canada's *Financial Post* estimated that in 1990, foreign investment in privately held real estate in Canada nearly tripled from the 1985 figure of U.S.$1.2 billion; if the debt portion (bank financing) of the real-estate transactions were included, the total investments of 1990 would exceed U.S.$13 billion.[85]

Vancouver was a particular favorite on the global shopping scene. A real-estate broker at Goddard and Smith Realty, Ltd., said of the real-estate market in Vancouver in late 1989: "The saturation point for Hong Kong investment in Vancouver real estate hasn't been reached yet because people now look at this city in a global context, whereas ten years ago only western Canadians were considered potential purchasers."[86] Hong Kong immigrants in the investor category were responsible for a large portion of real-estate development in Vancouver, despite the effort by the provincial government to channel the required capital into manufacturing, trade, and research and development. Investment categories acceptable to the B.C. government under the Business Immigration Program included "developmental real estate," defined as real estate to which substantive improvements would be made to carry on a business of significant economic benefit. In practice, this meant investment in virtually all commercial buildings, including office buildings, hotels, tourist ventures, and shopping malls.[87]

Investments in Vancouver real estate were aided by a number of consulting firms which, like the investor funds, sprang up following the initiation of the immigrant-investor category in 1986. Firms such as Perfect Coins International branched off from parent companies to better control the growing real-estate portion of the corporation's international activity. Often the parent company was initially involved in a financial connection between cities such as Vancouver and Hong Kong, the president of Perfect Coins told me in our August 1991 interview in Hong Kong.[88] Each real-estate project was targeted for the company's specific customers. The company did extensive research, chose one or two Vancouver projects, then marketed them heavily in Hong Kong, providing in-depth information about the property as well. Ancillary services such as banking, property maintenance, and information allowed the investor to make a good investment from a distance, and to maintain that distance if so desired. According to the president of Perfect Coins, this type of investment was almost completely speculative.[89]

## Urban Speculation and the Racialization of Restructuring

How did this global real-estate speculation become tied so directly to Hong Kong Chinese investment? In addition to the commercial real-estate purchases and megaproject developments, there was a massive influx of capital and subsequent redevelopment of residential housing in Vancouver's west-side neighborhoods.[90] These neighborhoods were historically among the most protected and exclusive in the city. Although there are no precise figures on Hong Kong investment in west-side property, a number of studies indicate extensive involvement in both purchases and development in these areas.[91] More important for the purposes of this study, however, is the *perception* of this speculative activity by long-term residents. This perception greatly exacerbated existing racism and contributed to the racialization of space and the ongoing struggles over spatial hegemony in the city.

One of the greatest sources of unease among west-side residents was the presale marketing in Hong Kong of newly built Vancouver condominiums. The Hong Kong–based firms Chi Wo Properties, Cheung Kong Properties, and Grand Adex were all involved in the presale marketing in Hong Kong of Vancouver condominiums, which captured approximately 60 percent of the retail end use. (For some condominium developments, this figure was much higher.) Local Vancouver residents were enraged that housing constructed in "their" city could be marketed first in Hong Kong—occasionally so successfully that the entire development would be sold out before reaching the Vancouver market.[92]

Information about these firms and their real-estate activities was available through a number of venues. Real-estate advertisements and brochures about the new luxury condominiums, for example, appeared in local papers such as the *West Side Weekly*. These ads showed prices averaging upward of C$600,000 for each luxury condominium, with penthouse suites for sale at well over C$1 million. For each condominium unit on the market, the entry into an exclusive, sophisticated way of life "in the very heart of Kerrisdale, the prime apartment location in Western Canada," was an advertised attraction.[93] For the Claridge development at 5850 Balsam Street (the site of angry demonstrations by evicted tenants), the real-estate firm Royal LePage advertised three penthouse condominiums as offering "what is unquestionably Kerrisdale's ultimate in luxury living."[94]

The expansion of Hong Kong real-estate networks in Vancouver's west-side areas became even more visible through papers such as the *West Side Real Estate Weekly*, delivered free to many west-side homes. In the October 27, 1989, issue, an increasingly common style of advertisement solicited property for marketing in Hong Kong: "Realty World in Hong Kong. Realty World–Kerrisdale will be represented by Lars Mogensen in Hong Kong at the Canadian Living and Housing Exposition, November 12–20, 1989. If you have been thinking of selling your property, here is a great way to get exposure. For more information and a free market evaluation, call ——." Bolder advertisements shouted: "Show your home to Hong Kong! Get your money's worth. List now to display your home in Hong Kong"; and "Hong Kong Market! How would you like your home to be seen on the Hong Kong market!!"[95] For the recalcitrant few who continued to resist these ads, letters mailed to residents at their homes and addressed to "Dear Property Owner" solicited the sale of their property.

Other visible evidence of a growing Hong Kong presence was manifest in the rapid proliferation of new real-estate companies that specialized in Pacific Rim investment. A Taiwan-based company, International Pacific Properties, Inc., opened a real-estate office in Kerrisdale in late 1989 with the slogan "Seizing the Pacific Rim." The opening was immediately commented on by the local newspaper, which exposed the global links of the company with large Pacific Rim corporations such as Pacific Construction, Pacific Wire and Cable, Pacific Sogo Department Stores, and Mitsui Rehouse.[96] In addition to this kind of statistical and anecdotal evidence, several prominent articles on Hong Kong investment in Vancouver real estate appeared in the media in 1988. These stories focused on the connection between rising house prices, demolition of older apartment buildings in places such as Kerrisdale, the construction of luxury condos, and the increasing activity of Hong Kong investors and developers in real estate.[97] In early 1989, the *Vancouver Sun* commissioned a series of articles on Hong Kong by staff reporter Gillian Shaw. The first, on the February 18, 1989, front page, was headlined, "The Hong Kong Connection: How Asian Money Fuels Housing Market." In the article, which showed a photograph of an Asian couple with a child on the stairs of a C$1.28 million west-side home, the link between rising real-estate prices and Hong Kong investment was categorical: "Although there are no immigration statistics showing the actual percentage of Vancouver real estate going to off-shore buyers, citizenship declarations filed with Vancouver's land title office and sales figures from the B.C. Assessment Authority show that among offshore buyers in the first ten months of 1988, an average 85% were from Asia. The majority of those were from Hong Kong."

Shaw also noted that the new buyers in Vancouver were paying more than a million *in cash* for houses that would have sold for half that price a year earlier: "Prices are being driven up by Asian investors who are willing to pay a premium over the local market value." Finally, the rapidity and scale of the purchases were emphasized: "In some strata developments, such as one at 1020 Harwood, where 20 units were sold in May to Hong Kong buyers, entire blocks of strata units are sold at one time in Hong Kong."

This article was followed by others with a similar focus. Shaw's March 17, 1989, front-page article was headlined, "Computer Shopping for B.C. Property." In this story, Shaw discussed a new internationalized computer listing system called Global Listing Service, Inc., which was being set up in Hong Kong to allow investors and immigrants to buy houses and secure venture capital via computer. Shaw quoted Steele (the founder of the system): "Suppose somebody in Hong Kong wants to move to Vancouver. They can punch in B.C. Vancouver, their price range, the number of bedrooms they want, and all the listing information about houses fitting their requirements will come up on the screen. They won't even have to go to Vancouver to buy."

Further articles in the series, on March 18, 20, and 21, 1989, were titled "Investment Anger Confuses Hong Kong," "Flipping Is a Hong Kong Game," and "Money Is King in Hong Kong: Entrepreneurs Find Paradise in the Streets of Hong Kong." In the article on flipping—buying and selling property for speculative purposes—there was a discussion of the presale marketing of Vancouver condos in Hong Kong, with an emphasis on the rapidity of turnover when housing is bought and sold for speculation.[98] In the article on money and entrepreneurship Shaw quoted several businesspeople and residents who stressed the emphasis on money making in Hong Kong. "Everything is geared towards money. There is no regulatory process here. . . . Here it is a different culture. It is geared to making money and there are few restraints on that. . . . The heroes here are the tycoons. Money buys respect, power and considerable influence."[99] In all the stories, the focus on the extremely wealthy in Hong Kong and on the general materialism of the culture produced a lopsided vision of the Hong Kong Chinese that resonated with and encouraged the common perceptions held by many white residents of Vancouver. Although the content of the articles reflected partial truths about Hong Kong society and the impact of Hong Kong investment in Vancouver, the headlines and general tone of the stories were highly inflammatory, as was the choice of topics.

These articles, which made explicit links between the rapid and unpopular changes occurring in west-side neighborhoods and increasing Hong Kong investment, exacerbated citywide racism. The conservative columnist Doug Collins said in an interview: "It's a funny thing about this country. We are willing to save the whales, but not the ordinary Canadians. . . . I think that the appointment of [B.C. lieutenant governor David] Lam was a signal to the Oriental world that we're ready to be taken over."[100] In our interview in Vancouver in October 1990, a coordinator at SUCCESS (United Chinese Community Enrichment Services Society) said of the atmosphere in the city in the late 1980s: "The reception here has changed. I guess that now some people feel that there are too many Chinese people, too many Asian immigrants. . . . And that's why we feel that the racial tension is there." Arthur Lee, a third-generation Canadian of Chinese descent, said of the mood of the city in mid-1989: "There was a bit of hysteria and paranoia. There was nothing overt—no demonstrations, nothing violent. That would be very un-Canadian. But the flap was ominous, and the feelings are still there."[101]

Callers to phone-in radio shows expressed fears of a Hong Kong takeover of the city's businesses and neighborhoods. New items appeared in stores, such as T-shirts with the logo "Hong-couver." Areas popular with Chinese residents were nicknamed Hong Harbour or Hongkong Row.[102] Donald Gutstein's 1990 book *The New Landlords: Asian Investment in Canadian Real Estate* sold out in Vancouver bookstores immediately following its publication. City Councilor Libby Davies received calls from people who blamed the changes in their neighborhoods on the Chinese immigrants. She said in our interview in Vancouver in February 1991: "I did get racist phone calls, particularly from older, white people—people who were born in Vancouver and from the west-side, who would say, 'Goddamn Chinese, it's the Chinese moving into our neighborhood.' "

Reporters from around the world picked up the story of increasing racism against the Hong Kong Chinese in Vancouver and linked it to the new immigration and investment programs that encouraged the wealthy to settle in Canada. On May 9, 1989, the *International Herald Tribune* headlined a story by Timothy Egan that highlighted Vancouver's racial problems "Fissures on the Pacific Coast: Asia Money Pours into North America, Stirring Unease." On June 4, the *London Sunday Telegraph* published an article headlined "Hong Kong Yacht People Buy Up Vancouver." In the first paragraph, the British reporter Peter Taylor's dismay at the immigration of wealthy Hong Kong investors into Canada was apparent: "If a Statue of Liberty was to be erected in Vancouver's magnificent harbour to celebrate the latest wave of immigrants from Hong Kong, the inscription might read: 'Bring me your property speculators and your fat cats, your huddled masses yearning to make a fast buck.' "

Hong Kong newspapers carried numerous stories on the increasing racism against the Chinese in Vancouver. Again, the link between the growing incidents of racial antagonism and the increasing movement of people and capital from Hong Kong to Vancouver was explicit. In the *Hong Kong Standard* for May 20,

1989, an article by Ken MacQueen entitled "Chuppies Become Latest Target of Vancouver's Racism" noted that immigrants and investors faced a mixed welcome: "part enthusiasm for their money, their education and their connections, part barely suppressed hostility." In the *South China Morning Post,* several articles discussed the lifestyle changes necessary for Vancouver residents in the face of increasing Hong Kong competition, particularly in the field of real estate:

> Laid-back and smug Vancouverites have been kicked out of their patio recliners by Hongkongers who don't waste a second when they decide it's time to move on an investment. Vancouverites now are being forced to do things Hongkong-style. Want a place to live? No more can residents afford to doddle around the city picking and choosing just the right place. The new reality is that you buy the first home that meets your needs as soon as possible—before a fax from Hongkong arrives.[103]

In late October 1989, in front of 5850 Balsam Street, a crowd of elderly tenants marched, sat, and hectored construction workers for several hours. This was the seventh in a series of demonstrations aimed at curtailing apartment demolitions in Kerrisdale. At all these demonstrations it was senior citizens, mainly women, who blocked bulldozers, occupied apartment units, and staged sit-ins, wearing signs that expressed anger about the loss of their communities. "Good Buy Kerrisdale$" and "Money First People Last" manifested the connections residents made between a declining sense of community and the unconstrained pursuit of profit evident in real-estate speculation and development.[104]

In contrast with earlier eras, the profitability of spatial reconstruction in these communities no longer benefited all who lived there. Many elderly women, forced out by rising prices as well as demolitions, were left without places to live. Betty Wright told reporters as she picketed the Balsam site: "I'm 80 years old and damn it all, I've lived in Kerrisdale all my life. I've had my name in for two years to get into a government-subsidized building and I've heard nothing. There is no place to go."[105] The unfortunate situation of these women, many of whom were evicted after living in a single apartment for more than twenty years, became the rallying point for activists all across the city. Several formed the

Concerned Citizens for Affordable Housing (CCAH), and tenants' rights groups complained vociferously about the rising displacement of the city's most vulnerable residents.

At one protest, five seniors stood in front of a doomed Kerrisdale apartment block and refused to move until they had expressed their concerns to a development representative. As time passed and the situation worsened, sympathy for the elderly Kerrisdale women grew. In a number of media statements and in private, city dwellers claimed that this case exemplified the increasing disregard for human relations evident in the drive for profit. The idea of human relations was always spatially positioned, transparently couched in the language of community membership. The normative equation of community belonging and the right to make a profit from that community was thus also greatly disrupted by the changes associated with the Hong Kong investment in the area. Spokesperson Annie Humphreys said bitterly at the time: "The people who belong to this city, this country, don't count anymore."[106]

The Kerrisdale situation occurred as a result of an old zoning amendment that allowed apartment construction in the neighborhood. To accommodate a housing shortage, zoning regulations in the 1950s had allowed multiple dwellings in a section of Kerrisdale. These regulations remained unaltered through the 1980s. In the frenzied, highly speculative market of that decade, developers seized the opportunity to demolish the old three-story walkups in the distinguished neighborhood, and construct twelve-story luxury condominiums for an upper-end clientele. What was particularly shocking for the older residents of this west-side area was the realization that for the first time in the history of the neighborhood, zoning was working *against* the firmly established values of old Point Grey. In our February 1991 interview, Committee of Progressive Electors (COPE) City Councilor Libby Davies said of the old regulations:

> What happened in Kerrisdale was that we had this very old zoning that had been there since the fifties or sixties, an old apartment zoning. Historically zoning has become more sophisticated and complex as society demands that it become more sensitive to various things. The Kerrisdale zoning for apartments was very basic, and it was wide open. In fact, it allowed you to build twelve stories under that old zoning.

Developers are smart—they looked around the city: clearly it was the west side at that time that was very desirable. They looked at this zoning, they realized it was what's called an outright use, they didn't have to go through a whole bunch of bureaucratic red tape at city hall to get a new development approved; they marched in there, bought those sites, and knew that they could turn low-rise fifties buildings into twelve-story, one-per-floor condominiums. And that is exactly what they did. The city reacted sort of minimally at the end of the crisis and put the damper on in terms of a zoning response, but it was pretty minimal. It was already well under way.

The demolition frenzy peaked in late 1989, when seven Kerrisdale apartment buildings were demolished in just seven months, and seventeen more buildings were scheduled to be destroyed pending City Council approval. The Kerrisdale demolitions were part of a major rise in demolitions citywide.[107] At the same time that numerous older buildings were razed, new residential construction increased. The value of annual building permits in residential areas during the 1980s shows a general increase between 1986 and 1988, and a marked increase between 1988 and the first quarter of 1990, when the luxury-housing market was at its peak.

Although residential building activity was high, the crisis in affordable housing worsened throughout the city. Most of the residential building took place in the luxury-housing sector, particularly in west-side neighborhoods such as Kerrisdale. The condominiums were priced on a significantly higher scale than the rental units, placing them well out of the range of the former tenants. Furthermore, although the condominium buildings were taller than the apartment buildings they replaced, the number of individual units available for rent or for sale was far fewer than in the former three-story walkups.[108]

The evictions from the original three-story apartments in Kerrisdale affected elderly white middle- and upper-middle-class women the most. Of the 3,225 renters in the area, 1,595 were sixty-five or older, most living in low-rise apartments constructed in the 1950s. Many of these women had sold their Kerrisdale homes several years earlier and moved to apartments in the area when they became widowed. As a result of their connections, race, and class, the women were considered honorary homeowners in the area, and their eviction and displacement was confusing and disturbing for all who ascribed to the community's values.

In our interview in Vancouver on October 6, 1990, Paul Smith, an organizer at the Downtown Eastside Residents' Association (DERA), said of Kerrisdale's blues:

> You know, in Kerrisdale, that's where the class gets all mushed up because you have someone who can afford it, who has a fairly good life, who has a bit of a pension and sees themselves as part of the upper class and then retires in their place, and all of a sudden they get evicted and they can't see a place where they can move into. And certainly women are in a worse situation than men are when it comes to this. In fact, most of the people I deal with out there seem to be elderly women. They're the ones who don't have the bigger income and rely on their old-age pension. They think of themselves as upper class and yet can't afford upper class anymore.

For many white Vancouverites, the elderly Kerrisdale women symbolized the negative effects of foreign investment and spatial restructuring, especially with respect to the fragmentation of local urban community. Their displacement provided a rationale for the necessity to harness "Asian" development and regain local control. Displaced white women, as honorary homeowners with acceptable values and seemingly undeniable rights to neighborhood residence, were used as evidence of the negative ramifications of Hong Kong real-estate speculation. The effectiveness of this representation allowed localists to assume the moral high ground from developers in the escalating battle over ideological control.

Rapid, dislocating development had occurred in the formerly elite and protected West End neighborhood at the turn of the century, but most of the upper-class residents of that area were able to make a profit from the rising house prices and relocate to the newly established and even more exclusive community of Shaughnessy Heights.[109] (See Map 1.) Raw land, planned and controlled by the Canadian Pacific Railroad, was still available for development and consumption. But for the elderly white women of Kerrisdale, land in Vancouver was fully developed, vacancy rates were near zero, their pension funds and dividends were no longer sufficient to make an upward or even a parallel move, and the venture capital involved in residential real estate was no longer completely controlled by local white capitalists with an interest in marketing a British place identity. These women subscribed to the tastes and values of Vancouver's white upper class, but

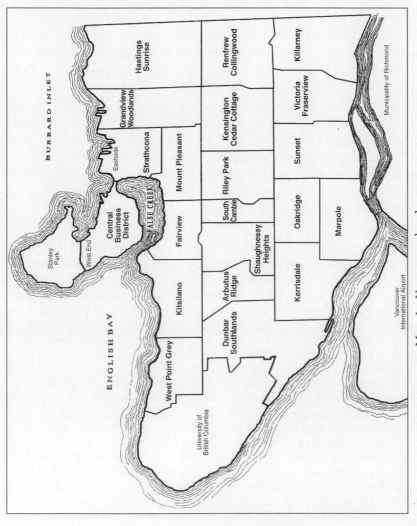

MAP 1. Vancouver local areas.

because of timing (the fact that they had sold their houses years earlier and could not make a substantial profit on recent price increases) and space, they were no longer shielded from the ravages of an intense and increasingly "vagabond" capitalism.[110]

The shock of this development in Kerrisdale galvanized many to take action. The Concerned Citizens for Affordable Housing (CCAH) was formed in 1989 after two packed Kerrisdale community meetings. (Approximately five hundred people attended the February 25 meeting.) According to Annie Humphreys, one of the CCAH founders, the large turnout reflected the scale of shock and dismay at the recent transformations. She told me in our October 1990 interview in Vancouver: "We're basically genteel people until something really pushes us that is grossly unfair. I don't think you've ever seen protests in Kerrisdale before. In Kerrisdale's whole existence! So things have to be pretty bad before you get west siders out." Councilor Davies in our February 1991 interview said of one of these early meetings:

> I remember going to the first meeting at the Kerrisdale Community Center, it must have been February 1989, and what kick started the whole thing and raised it into the public consciousness was the eviction notices for seniors in apartments in Kerrisdale. . . . I went to that meeting and it was overflow; there were more than six hundred people there. I mean, that's almost unheard of in Kerrisdale. And there were all these little old ladies screaming at the council members and provincial people saying, "You've betrayed us. You promised us stability, you promised us neighborhood stability and you've betrayed us." That was the message. It's something that was very, very different than what we've experienced in the downtown east side in terms of the public reaction, the media reaction, the reaction from elected officials. And I have no doubt in my mind it's because of where the problem was. And from that point on, I would say that that meeting in Kerrisdale was one of the key milestones, because it really leapt into the public consciousness. And it was this whole image of elderly women—very nice, white, middle-class ladies who were so self-sufficient and had never asked anybody for a dime their whole life, right—and all of a sudden they're at the mercy of the city and they're being exploited by these landlords and owners because of redevelopment. I think it sent shock waves through Vancouver; it really did. It was incredible.

Part of the unspoken pact between Kerrisdale residents and the pro-growth Non-Partisan Association (NPA) members of the City Council was the protection of west-side single-family neighborhoods. This protection was offered by NPA councilors (many of whom lived on the west side) in exchange for the powerful allegiance of these residents in voting behavior and in the general support of a primarily prodevelopment agenda for the rest of the city. The demolitions and dislocations in these neighborhoods in the late 1980s were thus both a shock and a betrayal. The betrayal of the Canadian dream was so profound and so unexpected that many tenants who had been displaced by the demolitions felt the zoning regulations that allowed these changes were some kind of a mistake. They did not believe that their exclusive, villagelike community could have ever been the site of development speculation in the past, or improper (not sufficiently protective and exclusive) zoning in the present.

The angst of the seniors was portrayed in a number of local and national newspapers. Although dislocations were occurring throughout Vancouver as a result of the affordable housing shortage and extremely low vacancy rates, the plight of these elderly women caught the attention of the media and municipal politicians. Human-interest stories on the negative experience of the demolitions for many of the women were commissioned frequently, with headlines that emphasized the shock of the evictions and of Kerrisdale's rapid transformation.[111] For example, "Bulldozer Blues," a December 15, 1989, *Vancouver Sun* article by Donna Anderson, profiled two elderly women. Both had been evicted from Forty-fourth Street. Hazel Dingwall, age seventy-five, was described as "miss[ing] her friends and lonely." Essie Goldberg, age eight-two, missed her community because it had "catered to seniors" with a community center and daily programs for the elderly. When she shopped in Kerrisdale, she had been on a "first-name basis with storekeepers."

The expressions of loss and alienation were frequent and evoked a general angst in the city that the media were eager to capture. Although these human-interest stories focused on demolitions and dislocations, particularly on the west side, they were juxtaposed with feature articles that chronicled the "tide" of Asian investment in west-side real estate. The connections between the dislocation of the women of Kerrisdale and the investment of "Asian" businessmen circulated in stories and anecdotes. One activist from the West End Tenants' Rights Association told me in an interview in November 1990 that "Kerrisdale was always on the capitalistic side. They have never voted NDP or COPE. Because they were always well off there. But, you see, it was the Asian investment that came in, they saw this Kerrisdale as a very good area to invest. And that's when they started to tear down these buildings. And these old ladies who had lived in this area for many, many years, well, they've been the first ones who have been kicked out of the neighborhood."

The rapidity of change, the anxiety about competition from an outside source, and the barrage of media stories about money, speculation, speed, and the internal connectedness of Hong Kong business practices added an extra layer of antagonism that permeated much of the society. Although the resistance to the demolition of apartments and the construction of luxury condominiums in Kerrisdale was framed rhetorically as a resistance to a loss of local control over land use and architectural design, the racial context of the struggle was indisputable. Individuals and groups who fought against the unwanted changes in their neighborhood often spoke of the battle as one of local control versus foreign ownership and international development. Left unsaid was the belief that these problems were caused not just by international capital, but by the fact that the capitalists involved in the changes were from Hong Kong.

Early Kerrisdale community meetings, for example, "quickly turned into a massive community protest for residents who blame Hong Kong money and land flippers for the drastic changes to their neighborhoods."[112] Likewise, early demonstrations against apartment demolitions directly questioned the negative effects of Hong Kong investment in the neighborhood.[113] Valerie, a Kerrisdale homeowner who had observed one of these early demonstrations, reluctantly admitted to me in an October 1990 interview the racial element of the resistance:

> V: I was in Kerrisdale one day and I went to a shop and along that road there was a house being taken away and people were there with placards and just standing quietly. And trees were being removed and. . . .
>
> K: What were the placards saying?
>
> V: I hate to tell you. I don't want to say. There's going to be upset. . . .

K:  Were the sentiments racist?

V:  Yes.

K:  Chinese go home? That kind of thing?

V:  Yes.

Despite the early anti–Hong Kong sentiment, the CCAH mobilized publicly behind the theme of affordable housing. Its agenda included ending the apartment demolitions, providing housing for seniors, and protecting neighborhood character. Annie Humphreys, as we have seen, was a prominent spokesperson and activist for the group. Her own mixed feelings about Asian investment, however, reflected some of the divisive elements involved in the group's formation, goals, and public stance. Paul Smith of DERA, in our October 1990 interview, said of her community work:

> [Annie Humphreys] is still quite active in Kerrisdale. She'll talk about foreign investment, much to the consternation of other people in the community and, in fact, in some meetings she's been shouted down because she does bring a spectre of racism and she has lost a lot of friends because of that. So you can maybe talk to her about that. Everyone else—it comes down to a lot of racist-type issues, and people just stay away from it. It's easier to talk about foreign investment or just developers coming in and being allowed to do what they want. Big capital.

In an interview with John and Jean Simmons of the Granville-Woodlands Property Owners Association, another grassroots neighborhood movement, I asked if their struggle against neighborhood transformation was primarily focused on changes in the character of the neighborhood or on the possibility that the new people coming into the area would be Asian. The difference of opinion between them reflected some of the ambiguities and ambivalence that public discussion of the issue often invoked.

> JEAN:  Well, the media tried to play that up, . . . that it was racist. But it wasn't at all. I mean, I don't care who's living where. It's the fact that the developers were coming in and just building these huge houses and they were building the huge ones because they did think that's what they wanted, . . . this opulence and size . . . because most of them came from Hong Kong.
>
> JOHN:  That's not true.

JEAN: Well, the Taiwanese are a big influx now too. It's just like they're in play land because they come from such small quarters and they come here and it's just what they want and —

JOHN: That's been a major hurdle all through this. The minute you start talking about what's happening to the neighborhoods in terms of the houses and the trees, they go, "Oh, you're a racist." I say, "No, I'm not racist." I don't care who lives here. Our block is multicultural.

The efforts of the CCAH and John Simmons to represent their slow-growth movements as resistance to unruly development rather than to Asian investment or Chinese people were derided by developers and politicians, who were able to point to many examples of racism throughout the city. The developers' efforts to win legitimacy by regaining the moral high ground were greatly aided by the establishment of patently racist organizations such as Residents Save Vancouver Please (RSVP), which were only poorly camouflaged as preservation movements.

The founding member of RSVP said in an interview with a local paper that the group was established to "deal with a number of symptoms of a hot real estate market."[114] When discussing the priorities of the organization, she listed investment controls, the preservation of the character of the neighborhoods, the prohibition of tree removals, and the protection of good rental housing against demolition. In a July 5, 1989, letter to the editor of *Western News*, however, this same RSVP member repeated the inflammatory lines that had appeared in the *Sunday Telegraph* just a month earlier: "If a Statue of Liberty was to be erected in Vancouver's magnificent harbour to celebrate the latest wave of immigrants from Hong Kong, the inscription might read: 'bring me your property speculators and your fat cats, your huddled masses yearning to make a fast buck.'" A *Vancouver Sun* reporter whom I interviewed in January 1991, and who had met with this woman and the other cofounder, believed the organization was racist and aggressive off the record but operated a "slick media relations campaign" that brought them a certain amount of publicity. The cofounder had been a local real-estate broker who was laid off as business activity with Hong Kong increased.

The indisputable evidence of racist agendas in many of the west-side social movements, and the use of euphemisms to camouflage these agendas, made any criticism of international capitalism or development appear suspect. Businesspeople and government officials interested in the unhindered integration of global capitalism could thus allege racism to silence the localists who called for greater controls on capital flow. Councilor Libby Davies was caught in this particular web. In our February 1991 interview, she said of the situation:

> Racism is alive and healthy in this city, unfortunately. Since its inception, Vancouver has been a multicultural city; this city was built by people of color. There are racists, and unfortunately this whole issue of development and of developers became characterized as a racist issue. And I think that was a wrong characterization, because it's really—I think the way to deal with it is an issue of capital. It's irrelevant where it comes from; ... well, I suppose whether it's national or international, there are some differences ... but it was really an issue of international capital that was flowing into the city. It's correct that a lot of it happened to be coming from Asia, from Hong Kong specifically, so they became the whole scapegoat for this. And it was pretty ugly. The whole thing became so much hypocrisy because in Campbell's defense, to unscramble himself from this, he hooked his hat on the antiracist thing, and anyone who complained was simply a racist. What he did was really to make people much more angry. I did get racist phone calls—particularly from older white people. People who were born in Vancouver and from the west side, who would say, "goddamn Chinese, it's the Chinese moving into our neighborhood." And it was sometimes difficult to explain to people that it was being driven by market forces as to what developers were building and tearing down. It's not that Chinese people exclusively wanted to have monster houses—that's what was available to them. Yeah, they did want to have bigger houses, but a lot of younger families want to have bigger houses, not just Chinese people. The whole racism thing became very much the scapegoat and it kind of worked both ways. Progressive people also got caught by it. Because we got characterized that way as well.

## Conclusion

The privatization of land and the heavy marketing of Vancouver as a new gateway city to the Pacific Rim spurred a massive influx of global capital in the 1980s. At the same time, new immigration programs and a concerted effort by the state to attract Hong Kong investment and entrepreneurs led to a major increase in the immigration of wealthy Chinese businesspeople and their families. Many of these Hong Kong immigrants invested in both commercial and residential real estate in the city, attracting even further investment from other global and national players. The overall economic and social organization of Vancouver changed drastically as a result. In contesting these changes, long-term urban residents drew on a narrative of the livable and humane city benevolently regulated by the state, where the reckless processes and agents of capital accumulation should be tempered in the interests of sustaining the beloved urban and national community. This liberal narrative, however, was contested by a coalition of developers, politicians, and investors, including some of the recent immigrants, who claimed that it was suspect on the basis of its fundamentally exclusionary territorial impulses.

The struggle over urban development in Vancouver thus forced a renegotiation of the assumptions developed over the past decades of an interventionist social liberalism based on the ideology of an egalitarian, humane, and livable city. This renegotiation was bound up with definitions of race, place, and culture that had far-reaching implications for municipal and provincial politics and for the spatial restructuring of the city. The ensuing debates about social liberalism and urban and national identity in Vancouver became saturated with references to race and racism, as well as to capitalism. Controlling racial definitions and explanations thus became a key tactic in the struggle to maintain the moral high ground in the battle over interpreting the changing urban landscape.

In the increasingly high-stakes battle over hegemonic production, the power to define race and racism was paralleled with an equally intense struggle over the power to determine when the issue of race and racism could become an acceptable topic of public discussion. The carefully demarcated lines in liberal theory between public and private were blurred and contested as differing actors and institutions, from builders to long-term residents to the media to the immigrants themselves, sought to influence public opinion and hegemonic discourse.

In public debates and images, a consistent theme promulgated by localists was that of elderly women as modern-day Davids battling the Goliath of global capital. This is a classic local-global theme related to globalization and the erasure of the values of (national) community and of the socially liberal goals of social reproduction. The well-publicized confrontation, however, contained an implicit subtheme that undermined the proponents of slow-growth development. That subtheme was an undercurrent of distress over a victorious, highly racialized Chinese entrepreneur, emblematic of global diasporic capital, usurping the spaces of the elderly white women, a metonym for community, belonging, and nation. This was an image of capital, motion, speed, and the nonwhite infiltrating the pure, white, fixed contours of the beloved Canadian landscape. Even as an undercurrent, this atmospheric mood (greatly exacerbated by the media) was strong enough to make resistance to the transformations in Vancouver widely, even globally, understood as having a racist caste.

In his discussion of liberalism, empire, and territory, Mehta dwells on the absence of a notion of "territory" in Lockean conceptions of political identification and membership. Classical liberal thinkers denied any link between the constitution of space and the constitution of society, despite the clear territorial identification of most individuals with a particular place and the valorization and defense of its borders.[115] Despite the allegiance of most societies to the territories with which they are identified, however, even contemporary liberal thinkers have, for the most part, denied the linkages between the formation of a political identity and the constitution of territory. This neglect has important ramifications in *neo*liberal discourse, which similarly erases the significance of this link in its imagining of a clear, smooth, and level playing field at the global scale, at the same time associating this erasure with freedom from historical, national, and community-level frictions and antagonisms.

Because of this divergence between the discourse of spacelessness and the actuality of territorial (in this case, urban) identification, many urban residents of Vancouver considered the neoliberal theory *and* practices of the state flawed and contestable. The offshore speculation in the housing market galvanized by provincial and urban policy, as well as by the federal Business Immigration Program, were critiqued on the grounds of their erasure of urban history and sentiment, in addition to their fundamental undermining of the socially liberal components of economic redistribution and equal entitlement to the city. In this way, the neoliberal agenda was contested on the basis of its violation of a "structure of feeling," in Raymond Williams's sense of an authentic cultural community grounded in the symbols and practices of everyday life.[116]

In this scenario, however, the threatened structure of feeling could, in some instances, be portrayed as a highly differentiated urban landscape striated with racial and class hierarchies formed and maintained through a British-inflected set of cultural distinctions. Thus, while long-term urban residents could attack a neoliberal urban agenda for its mercenary motives and lack of human warmth or sense of territorial allegiance, promoters could uphold it as a global agenda untainted by local, racial, or class antagonisms—unlike the spatially sedimented world of actually existing urban neighborhoods. This type of discourse was difficult for social liberals to deflect, as it exposed the hypocrisies of a liberalism that similarly denied the importance of space in the abstract, but nevertheless relied on an implicit sense of territorial allegiance to delimit authenticity and belonging within the community.

# 3 The Spatial Logic and Limits of Multiculturalism

THE HISTORICAL CONTEXT of Canadian confederation and the geographical distinctiveness of Québec's role in Canada are crucial for understanding the national development, interpretation, and implementation of both liberalism and neoliberalism on the ground, and especially their incarnations in concepts such as multiculturalism. Multiculturalism emerged as a form of social liberalism in Canada during the 1960s and 1970s, but its historical provenance stretches back much further. The recognition and embrace of "difference," the philosophical core of contemporary multiculturalism, was a necessary strategy for national unification as far back as the time of federalism and the development of the modern Canadian constitutional system under the British North America Act of 1867. Earlier, a deliberate effort to "submerge and obliterate the distinctive nationality of French Canada within the framework of a united colony" had led to growing internal dissension and an increasing sense of nationalism among French Canadians. The frictions and blockages caused by this British-French antagonism damaged the incipient political system of the colony and "hindered economic expansion."[1] Tolerance and the acceptance of difference were necessary for both modern state formation and capital accumulation in Canada from the very beginning of colonial settlement.[2]

The development of a pluralist ethic thus took shape within a particular spatial form: a territorial state divided between two often antagonistic colonial factions, tenuously held together by mutual economic advantage and a common distrust of their southern neighbor. Later iterations of cultural pluralism, and then multiculturalism, grew out of this history and geography, and involved a strong demarcation from the philosophy and practices of the United States.[3] The development of liberal thought and practice in Canada, and the manner in which Canadians produced and used liberal imaginings, thus reflected the positioning of the two internal "nations" within the state, as well as the overall position of the state within a larger structure of international relations. In short, multiculturalism in Canada was produced through and reflected in a highly specific geographical logic of the territorial state.

In recent years, however, this territorial logic has been shifting in reaction to the twin processes of state deterritorialization and neoliberalism. A new spatial formation based on *transna*tional connections and globally networked affiliations between people and places is creating a different kind of relation between the state and capital, and between the state and its citizenry. This has led to a shift from a multiculturalism directly implicated in both state formation and capital accumulation to a more strategic and entrepreneurial global cosmopolitanism. Contemporary neoliberal policy is oriented to the global accumulation and circulation of capital; the multicultural emphasis of state politicians and bureaucrats now focuses on the need to harness partnerships of difference *across* national borders.

Although multiculturalism was always bound up with the development of capitalism, it was tied to an ideology of capital accumulation for the development of the nation. Alongside this prominent discourse was the narrative of what I term "the multicultural self." The multicultural self was part of a broader social component of liberalism, which sought to inculcate a sense of tolerance as part of a citizen's obligation toward national social coherence. As a socially liberal philosophy and policy, Canadian multiculturalism invoked a complex mix of tolerance of difference, social equity, opportunity, and nationalism, with an underlying but fairly opaque history of capitalist articulation. In the neoliberal era of the late 1980s and 1990s, however, this liberal ethos was exposed as at best a superficial effort aimed mainly at placating the French so as to get on with (British) business as usual. At the same time, the tolerant Canadian national subject was undermined as an important position in favor of a strategic multiculturalism premised on international cosmopolitanism and global economic advantage.[4]

## THE LOGIC OF LIBERAL MULTICULTURALISM

One of the central tenets of multiculturalism in Canada, as else-where, was the idea that all groups have the right to claim recognition for their unique "identity" in the national public forum. Identity was generally conceived of as cultural identity, and it was this cultural identity, especially cultural difference, which the philosophy of multiculturalism allowed and encouraged. Despite its shifting form and multiple adaptations, a central logic under-pinned multiculturalism in Canada, premised on these normative assumptions of what constitutes identity, as well as what constitutes the liberal public sphere of the nation.

This logic was implicit and remained largely unrecognized by numerous commentators. However, with the contemporary transnational movements of people and ideas, this underlying logic became more evident, as did its contradictions and limitations. As a concept, multiculturalism relied on the notion that individuals and groups were entities with specific and particular *identities*, and that these identities could be *identified by* the state. When these identities were seen to be culturally distinct from each other (as they had to be, since that was one of the state's defining criteria of identity), the state should intervene to ensure that they were publicly recognized and treated equally. The entire notion of identity was thus predicated on a belief in autonomous individual personhood and state recognition of it. According to Roger Rouse, "the most obvious meaning of identity as a definition of personhood is that of a sameness or continuity of the self across time and space. But how is this continuity thought to be secured? The dominant view within the social sciences is based on the idea that identity is a kind of property, an idea manifested most clearly in the widespread tendency to describe identities as things that people have or possess, claim, acquire, lose and search for."[5]

Identity, in this dominant perception, was understood to "belong" to autonomous individuals; it was based on sameness over time, and also on its difference from other identities. Further, if particular properties intrinsic to identity did not continue across time and space, then the continuity of the self or of the collective (with the "collective" defined as a group of autonomous individuals) could be presumed to have been "lost."

As Rouse notes, this hegemonic discourse of identity is a relatively recent phenomenon with strong links to the historical rise of liberalism in Western societies. The idea of the rights-bearing individual with complete and autonomous proprietorship in the self can be linked to the works of the early liberal philosophers. They associated the right of the individual to self-possession with the relinquishing of certain other rights, as part of a "social contract" with the state. These rights also, as noted earlier, were linked with the emergence and development of capitalism as a rapidly expanding socioeconomic system in the seventeenth and eighteenth centuries.[6] Within this intellectual framework, individuals were free to form contracts, own property, and pursue their private visions of the good life without fear of interference from society, as long as they followed the dictates established by their multiple formal and informal agreements.

In many societies, however, this liberal intellectual heritage has not been as influential as in the West, and the individual and identity are understood in quite different terms. For example, in traditional rural Chinese society, as described by the sociologist Fei Xiaotong, the dominant understanding of personhood is not of discrete units coexisting equally and impersonally, but of members of extended families acting out their proper "roles" in a dense and hierarchical social network of obligations and expectations.[7] Similarly in Rouse's study of the peasant population of Aguililla, personhood was generally defined not as autonomy and self-possession, but as "the occupation of a particular place within a pre-existing field of social relations."[8] In these conceptualizations, the self operates within a collective, an organic whole, not as one unit out of a set of related yet discrete and independent parts. This more holistic, reciprocal, networked, and hierarchical understanding of the self in relation to the collective is common to many non-Western societies.[9]

In contrast with this spatially flexible, socially networked understanding of personhood and its recognition in society, the recognition of cultural difference in Canada was based on the necessity to recognize the individual or the group as a distinct and

discrete geographically locatable entity. The constitution of difference and the process of differentiation were based on singularities, on isolating and separating one set of cultural norms
or behaviors from another.[10] Further, in liberal states such as
Canada, the recognition of cultural difference took place in the
public sphere of state discourse; where recognition of a distinct
cultural community occurred, the nation was assumed to be its
de facto territorial container.[11] In most official rhetoric and policy associated with multiculturalism in Canada, culture and cultural difference were equated with community, and community
was assumed inherently locatable within the spaces of the nation.

The ways in which these multiple spatial assumptions came
together formed a logic of multiculturalism that was firmly
enfolded within Western liberalism and acted ceaselessly to perform the modern nation. But more than this, these assumptions
operated effectively and silently through time to contain and
domesticate the population, and to render normal and acceptable
Canadian state policy and intervention on issues such as immigration, Québecois "difference," conflicts over urban land governance, and the method and pace of integration into the global
economy.

The context of contemporary transnational migration from
Hong Kong, however, exposed several unsavory aspects of this
interventionist social liberalism. For example, as provincial and
federal politicians yoked official multicultural discourse to the
aims of expanding economic ties with Pacific Rim entrepreneurs,
the longstanding linkages between the growth and development
of liberalism and of capitalism were rendered much clearer. Multiculturalism, a key manifestation of Canadian liberal tolerance,
lost its nationally beneficial and neutral aura as it was strategically manipulated in the interests of multinational capitalism.[12]

In addition, the movement of the transnational migrants across
space disrupted the assumptions of a multicultural nation based
in a fixed territory.[13] Although enshrined as the official doctrine
of national coherence—as the social glue that held disparate
national members together in a single beloved "community"—
the coherence of most interest to politicians in the late 1980s and
1990s was the articulation of capitalisms. The state project of

managing diversity, a technology of control through domestication, was deterritorialized and thus denatured as state leaders channeled their multicultural desires far across the Pacific Ocean.

The process of domestication assumes a bordered space. In the case of multiculturalism, up until the 1980s, this space was assumed to be the territory of the state. Yet Canadian politicians, business leaders, and the Hong Kong immigrants themselves called on multiculturalism to stretch beyond the state to incorporate "ethnic" citizens and denizens associated with Canada wherever they might be found. Unsurprisingly, the newly deterritorialized understandings of multiculturalism led to a new set of sociospatial affiliations and loyalties. These included new divided loyalties to multiple states and nations, as well as loyalties to scales other than the state (regional, city, village, and supranational). Multiculturalism also began to incorporate the people who moved between scales at different moments in time, or who jumped scales to avoid precisely the types of regulation and domestication which state multiculturalism originally effected. As transnational migrants increasingly began to move in the interstices of the nation—sometimes deliberately, sometimes forcibly, and sometimes because of different subjective understandings *of* the nation—multiculturalism as national narrative and as tool of domestication waned dramatically. As there was no longer a "domestic" Canadian, technologies of domestication ceased to work. And as the constitution of a Canadian multicultural self waned in importance, the strategic global cosmopolitan rose.

Finally, the processes of transnationalism, broadly conceived, operated not just to disrupt the nationalist project of Canadian multiculturalism, but also to expose its inherently British framing. Despite the rhetoric of equality for the differing cultures of Canada, the variety of liberalism evident in the Canadian Charter reflected a British legacy of a rights-based, procedural individualism, rather than the more communitarian aims of collective cultural "survival" for French culture in Québec. This was one of the dominant motifs of the Meech Lake Accord.[14] Multiculturalism, first recognized as cultural pluralism in Canada, was intended to disguise the profound British centrism of Canadian liberal culture, especially with respect to French desires. But the dominant perception of "Canadianness" as equivalent to British cultural norms and practices became more and more apparent with the increasing clashes of "other" cultural norms and expectations—particularly those that refused to accede to an implicitly British cultural privileging.

Zizek has noted how this historical (socially liberal) multicultural whitewash also served to mask the racism inherent in the abstracted and distanced definition of the Other from the vantage point of European cultural superiority: "Multiculturalism involves patronizing Eurocentrist distance and/or respect for local cultures without roots in one's own particular culture. In other words, multiculturalism is a disavowed, inverted, self-referential form of racism, a 'racism with a distance'—it 'respects' the Other's identity, conceiving the Other as a self-enclosed 'authentic' community towards which he, the multiculturalist, maintains a distance rendered possible by his privileged universal position."[15]

As frictions around house styles and personal behavior increased in Vancouver in the late 1980s, various state institutions called on multiculturalism, not just in the recognition of immigrant difference, but also in the active inculcation of appropriate cultural behavior. Proper behavior was behavior that would not be deemed offensive to culturally British Canadians, and that would allow everyone to "get along." It was also behavior firmly locatable within the disciplining apparatuses of the state. The fundamental illiberalism of Canadian state liberalism was evident in its inability to digest a way of thinking or mode of behavior outside a British cultural framework.[16] Difference was acceptable, but only when it worked *for* the nation, within a clearly proscribed cultural framework. If multiculturalism clearly began to "mean business" in Canada in the 1980s, it also began to mean just getting along, British-style.

## CULTURAL PLURALISM, STATE FORMATION, AND CAPITAL ACCUMULATION

Canada's early colonial history was marked by divisions between the two charter groups, the French and the English, as well as by territorial struggles with the indigenous people. Finding common symbols and meanings of community and national identity was fraught with difficulty, as each group contested the unifying themes and values of the other.[17] Early efforts to increase a unified spirit of nationalism heightened British and French antagonisms and threatened to break the federal system apart. Yet both sides felt the growing threat of U.S. hegemony. The fear of being absorbed into the U.S. culture and economy promoted an active search for a symbolic vocabulary of national identity that continues today. Jack Curtis, a professor at the University of British Columbia in Vancouver, responded with concern in our June 1991 interview to my question about the issue of identity in Canada:

> The country is racked with anxiety. "Anxiety" is too gentle a word. We're worried sick. This country is a precarious experiment. We're virtually atop the enormous and in some ways aggressive ambitions of the United States. . . . And then this is an enormously regionalized country. Especially now when nation states are under assault anyway. It is a place that has really tried to grapple with difference. After all, it was created out of two fundamentally different peoples as well as the natives. Different cultures. Somehow one has to reconcile that. And there is a symbolic vocabulary to work out. Except for a few symbols like the maple leaf there isn't much in common. Even the beaver is rather a Protestant symbol. The place is really a coalition. A lot of people are committed to this experiment of variety. If you're looking for a single Canadian identity, there isn't one. It's all mixed together. It's very, very precarious.

The hazard of promoting symbols as seemingly innocuous as the beaver illustrates the difficulty of unifying and representing Canada as a nation. Cultural pluralism, heralded as Canada's answer to the U.S. melting-pot metaphor, appeared to many the only possible solution. Upholding citizens' equal rights under the law, yet respecting individual differences that stemmed from diverse cultural and "racial" backgrounds, seemed the perfect solution for Canada's disparate population.

The first person to promote cultural pluralism in Canada was Sir Wilfrid Laurier, Canada's first French prime minister.[18] At the time of his election in 1896, Canada was experiencing an economic boom, and Laurier saw his mission as one of investing Canada's soul and spirit in a "singular nationality" so that the country could expand and grow without the drag of conflicting sentiments from the two charter groups.[19] During the years Lau-

rier was in office, he consistently promoted national unity and
the reduction of animosity between the British and the French.
Despite his French heritage, Laurier's Canadian allegiance was
staunchly British. In a letter to a friend in 1909 he expressed his
commitment to a pluralist conception of nationhood, but also to
his assumed mandate as a "British" subject: "We are British subjects, but we are an autonomous nation; we are divided into
provinces, we are divided into races, and out of these confused
elements the man at the head of affairs has to sail the ship
onwards, and to do this safely it is not always the ideal policy
from the point of view of pure idealism which ought to prevail,
but the policy which can appeal on the whole to all sections of
the community."[20]

Laurier made his commitment to "all sections of the community" in a time of economic prosperity and political intrigue,
when the relationship between politicians and businessmen was
especially close. Although the preceding Macdonald government
was infamous for the tight relationships between early railway
promoters and cabinet members, the "Laurier Plutocracy" was
similarly endowed with "bankers, engineers, corporation lawyers,
railway builders, mining promoters, pulp and paper producers,
and public utility entrepreneurs."[21] It was a time when a select
group of elites, often holding simultaneous positions as politicians and the heads of corporations, "could correctly be called a
ruling class representing the dominant political, economic and
military forces of their time."[22]

The close links between businessmen and politicians at the
turn of the century were crucial for the development of the Canadian Pacific Railroad, the telegraph, and numerous personal fortunes. These fortunes were held almost entirely by individuals
and families of British descent, and were maintained by a white
Protestant monopoly in the banking industry that is still evident
in contemporary Canada.[23] British-born elites controlled the banking, insurance, communications, and transportation industries
and held most of the high government offices.

Members of this group tended to favor financial capital over
industrial capital; they crippled indigenous industrial development
by withholding capitalization of industrial projects. Commercial

capital was loaned primarily for resource extraction and trade that served the industrial development of the United States and Great Britain. Clement wrote of this investment in international money capital: "It is this international character of Canadian capitalism, unable to contend with trade restrictions and international fluctuations, which dominated Canadian political economy."[24]

Given the consolidation of economic and political power in finance, transportation, and services rather than in the industrial sector, political pressure was brought to bear on the promotion of policies favorable to commercial rather than industrial capital, including the free and unhindered East-West and international circulation of capital and goods. The circulation of capital and goods was facilitated greatly by good "race" relations between the charter groups; Laurier's professed commitment to a liberal democratic tradition of cultural pluralism thus dovetailed with his supporters' desires.[25] The rhetoric of national unity and racial harmony was propounded frequently, with the universalist concept of a plural and tolerant "Canadianness" declared the best strategy for both state formation and capital accumulation.

It was considered self-evident to the political elite that the notion of Canadianness applied only to the two charter groups. Indigenous groups were denied a voice in planning the country's future, and exclusionary immigration policies against the Chinese continued. Furthermore, despite the liberal rhetoric of tolerance and belonging, the concept of "Canadianness" was firmly grounded in British values and traditions. Those who felt that this version of cultural pluralism was not to their advantage (the French, for example) were attacked as driven by "racial particularism."[26]

The most obvious example of British power in defining the cultural meaning of Canadianness was manifested in the Boer War crisis. During the Anglo-Boer War (1899–1902), Canadian participation on the side of the British was considered inseparable from Canadian nationalism and patriotism, even though France was not directly involved in the overseas confrontation. French Canadians who opposed the war effort were vilified as unpatriotic and traitorous. Henri Bourassa, the nationalist leader of Québec and one of the most vociferous opponents of Laurier's wartime policies, was roundly condemned in the media and in England, while Laurier was hailed as "one of the enlightened French Canadians who fostered national development."[27]

Any French person who opposed the Boer War, Canada's naval policies (which supported the British), or cultural pluralism was labeled provincial and racialist. The assumption of British values and culture was the norm; all else was "tribalism." Those who opposed the norm were represented as un-Canadian and a hindrance to unified national development.

Sixty years later, many of these historical cultural norms remain ingrained in the rhetoric and practices of contemporary multiculturalism. Despite the official move by Prime Minister Trudeau away from a cultural pluralism that emphasized only British-French relations, the multicultural agenda, well into the 1980s, remained focused on French heritage and on placating French unease in the face of continuing British Canadian cultural dominance. Even with this emphasis, however, the core assumptions about authentic Canadianness remained firmly ensconced in a British set of cultural values, as did the fundamental workings of the economic world.

State policies that were developed to foster a wider multicultural ethic in Canada were fairly superficial, involving a few educational efforts to encourage acceptance of "alternate" cultural traditions. This acceptance rarely extended beyond ethnic multicultural fairs such as the Toronto Caravan, and in fact served to underline a "normal" Canadian culture against which all others were measured. More fundamental changes that involved economic redistribution or the recognition of group demands were not endorsed.

This period, like the Laurier era a half-century earlier, was thus characterized by a loud state rhetoric of recognition of difference, equal opportunity, and the importance of inculcating social tolerance *for* the nation. In some cases this rhetoric, implemented in educational systems and disseminated through the media, helped create a more tolerant state citizen.[28] But in terms of the actual redistribution of economic wealth or access to the corridors of power, liberal multiculturalism changed little in the 1960s and 1970s, as we will see.

## THE POVERTY OF CHOICE IN LIBERAL MULTICULTURALISM

They are easily stirred by bilingual cornflake boxes, a statement by General de Gaulle about a "free" Québec, or a proposal to establish another school in Ontario in which French is to be used as the language of instruction. Local skirmishes and battles are fought with passion over linguistic issues while the economic war is lost every day in the boardrooms of cities like Toronto and New York.

—Ron Wardhaugh, *Language and Nationhood*[29]

A major source of disquiet in Canada in the 1960s was the growing strife between the British and the French. French discontent was expressed in the early sixties in the "Quiet Revolution," the first major expression of French separatism in Canada.[30] The language of separatism and incidents of terrorism precipitated the 1963 Royal Commission on Bilingualism and Biculturalism, a study whose express purpose was to reduce friction between the two charter groups. This search for pluralist harmony, for an ideology that could cement the nation together during periods of crisis, was also a desperate search for national legitimacy. The link between multiculturalism and nationalism is a longstanding one. Rubinoff wrote in the early 1980s: "To believe in the values of any particular conception of pluralism is first of all to endorse a particular conception of nationalism." Similarly, Gupta and Ferguson noted a decade later: "'Multiculturalism' is both a feeble acknowledgement of the fact that cultures have lost their moorings in definite places and an attempt to subsume this plurality of cultures within the framework of a national identity."[31]

The report of the first Royal Commission on Bilingualism and Biculturalism included an apparent addendum, submitted to the governor-general in 1969. "The Cultural Contribution of Other Ethnic Groups" opposed cultural legitimacy for groups other than the first two colonial powers. Pluralism and Canadian nationalism in this report extended only to an inclusion of the cultural norms of the French and British:

> Many of the non-British, non-French groups accept bilingualism but categorically reject biculturalism. They concede Canada to be a country with two official languages, but argue that it is fundamentally multicultural. Against this view, the Commission strongly supports the

basic bicultural nature of our country referred to in its terms of reference. Although we should not overlook Canada's "cultural diversity,"
this should be done keeping in mind that there are two dominant cultures, French and English.[32]

Despite the views of the commission members, Prime Minister Pierre Trudeau outlined a new plan for the country in a 1971
speech, "Multiculturalism within a Bilingual Framework."[33]
Trudeau ended the speech with an image of Canadian identity
in a new era. This identity was predicated on "the conscious
support of individual freedom of choice," the freedom to choose
one's identity, to just "be ourselves." The word "freedom" is
also ubiquitous in a number of liberal speeches that define
Britishness in England.[34] In these conceptualizations, the idea of
freedom was limited to the rights of the individual to choose to
"make a contribution" and "to hold distinct personal views."[35]
These were rights, moreover, firmly guided by the strong but
fair hand of the state, as Trudeau made clear in his speech: "In
conclusion, I wish to emphasize the view of the Government
that a policy of multiculturalism within a bilingual framework
is basically the conscious support of individual freedom of choice.
We are free to be ourselves. But this cannot be left to chance. It
must be fostered and pursued actively. If freedom of choice is in
danger for some ethnic groups, it is in danger for all. It is the policy of this Government to eliminate any such danger and to safe-
guard this freedom."[36]

Trudeau's speech represented the preservation of ethnic identity as voluntary, which indicated the government's emphasis on
an individual's right "to choose" rather than on a group's collective rights or on a notion of personhood based on social roles
within a socially networked framework. Further, Canadian multicultural policy often defined an individual's rights in the negative, as the right *not to be* associated with an ethnic identity, or
the right *not to be* oppressed for association with an ethnic identity. The liberal democratic notion of individual equality of opportunity under the law was thus upheld but relegated to the private
sphere. In private, individuals of any ethnicity were accorded
equal societal opportunities regardless of cultural "difference." In
the public realm, however, collective rights remained entrenched
in English and French configurations. During this time, despite

much rhetoric to the contrary, attainment to any public position of power was predicated on "required acculturation to prevailing Anglo or Franco norms and practices."[37]

Following Trudeau's speech, multiculturalism was implemented as an official social philosophy in the areas of education, social welfare, and the law.[38] Multicultural policy was primarily focused on issues of lifestyle, or "expressive" elements of ethnic difference. Educating people about diversity and individual choice were the primary tools used to encourage a popular acceptance of cultural difference. The emphasis on cultural difference and prejudice, rather than on unequal access to resources or systemic relations of power, allowed the government to promote multiculturalism as a philosophy of individual self-improvement. Moreover, the constitution of this "multicultural self" was linked with the development of a stronger, more coherent, and more successful nation.

In the Toronto Caravan, an annual fair in Ontario where Canadian minority groups were invited to display their "ethnic" ways of life, for example, Canadian visitors were educated to appreciate and respect the cultural differences of immigrant groups. Although intended to increase tolerance and national cohesion, this form of "red boots multiculturalism"[39] emphasized the distinctions between "authentic" British Canadians and "the rest." Programs like the Toronto Caravan also encouraged what Porter called "the soft face of discrimination, ... the face of one who is charmed by cultural difference, who finds variation of usages an amusing natural display. Such a charming spectacle is produced by regarding cultural difference aesthetically, i.e., as an array of objects that is pleasing in its colour and its unusualness of form."[40]

The spectacle of multiculturalism in the Toronto Caravan was intended to bolster the idea of a unified and modern nation with a shared sense of national identity. Under the benevolent direction of the state, different ethnic groups were presented as harmoniously engaging in the performance of "unity within diversity." The multicultural ideology of the late 1960s and 1970s was predicated on a belief in the recognition and harmonious coexistence of different cultures. What was not supported in state policy, however, was the notion of a *structurally* pluralist society that would entail a reworking of economic structures or political organization.[41] In addition to the threat to established elites which this entailed, structural pluralism provided a basis for multi*nationhood*, anathema to the state. Kallen wrote of this possibility: "Should the mosaic take the form of pluralism in the public sector, then ethnocultural rights could be guaranteed through political representation, economic (occupational) control in specified areas, recognition of linguistic rights, and (in its most extreme form), territorial (regional/local) autonomy. The latter form of pluralism provides the basis for *nationhood* based on the geographical separation of ethnic collectivities sharing language, culture and territory."[42]

To avoid these types of spatial claims, the initial wording of the government's multicultural policies was deliberately ambiguous. Without fixed categories or meanings, state organizations disbursed funds for multicultural projects according to a disjointed and haphazard set of criteria. The allotment of funds thus became a means of state control, since the state could determine which minority ethnic organizations and which projects were legitimate. Some voices and demands were given the monetary resources that allowed them to be heard, while others were silenced through lack of funds.[43]

Multicultural policy was critiqued by many French politicians and scholars who felt that the importance of French culture as one of the two "founding" cultures in Canada was being subsumed by a mishmash of many "equal" cultures. They observed, further, that despite a rhetoric of multicultural equality, French culture was being thrown into the ethnic vat of multi-Otherness, while British culture was once again displayed as the norm against which all else would be measured.[44] Other critics elaborated on this theme, arguing that the mosaic imagery was predicated on an assumption that the liberal state would uphold the rights of Canadian citizens as individuals, *despite* difference. This assumption relied on an acceptance of the idea of a state, in the British liberal tradition, based on "a single set of legal and political principles to which all its citizens owe their allegiance and which form the basis of their patriotism and collective identity."[45] It also relied on a static conception of identity, wherein all ethnocultural collectivities were able and willing to maintain distinctiveness and boundaries, not dominate or surrender to other groups, and limit and control interaction with others to an acceptable level.[46]

Those who failed to acquiesce in these assumptions were
labeled tribalist, racialist, or, worst of all, unpatriotic. Being Canadian entailed implicit acceptance of a liberal British Canadian
way of life. This refrain was evident in the keynote address of
Member of the Legislative Assembly (MLA) Kim Campbell in a
1988 conference on multiculturalism and policing in British
Columbia. The minister underscored the *limits* of difference and
affirmed the role of the state in upholding the values of a common proceduralist and nationalist culture: "In our society, our law
and our society are built on the notion of social trust where the
instruments of the state act in the interests of the people. We can
have people in our courts who may have a different set of values,
whose loyalty to their family takes precedence over any loyalty
to a state for very good historical, cultural reasons. We must teach
and convey and communicate that that is an inappropriate value
in Canadian society."[47]

The freedom of choice in liberal multiculturalism was thus
constrained in a number of ways. It was freedom for the individual, not the group, and it was premised on adherence to a secular, liberal conception of the nation and on the values of an elite
way of life. British Canadian norms were maintained institu-
tionally and spatially, despite liberal rhetoric that guaranteed the
right to be different. At best, the freedom to be different was the
freedom to choose to dance at the Toronto Caravan or to eat gra-
nola out of bilingual cereal boxes. Anything more profound,
incorporating political or economic restructuring or territorially
defined concepts of multi*nationhood,* threatened the authority of
the state to define Canadianness and was vigorously attacked.

## RACE RELATIONS AND GLOBAL MULTICULTURALISM

The discourse and practice of multiculturalism have been integral to
the process of administrative normalization within the framework of
the . . . state. Because fundamentally *different* traditions are described
as necessarily *contradictory* (and therefore in need of regulation), state
power extends itself by treating them as norms to be incorporated and
coordinated.

    —Talal Asad, "Multiculturalism and British Identity"[48]

In its original manifestations, the idea of culture in Canadian
multicultural policy and rhetoric was fairly static and circum-

scribed, focusing on essential and essentialized differences and on generally superficial attempts to preserve the cultural heritage of minority groups, primarily the French. The early multicultural policy initiatives functioned as a relief valve for tensions with Québec and as a framework for a national discourse on the possible reconstitution of Canadian identity.[49] Multiculturalism focused primarily on questions of cultural heritage and national identity, with a strong educational rhetoric of tolerance and self-directed improvement in combating ethnic discrimination. The weakest policy programs were those targeted at reducing institutionalized racism. Aside from a few films and radio programs, little money or time was spent in this area.[50]

In the mid 1980s, however, this emphasis began to shift. The early concern about identity in the context of friction with Québec became more widespread and all-encompassing, and political and scholarly pronouncements about cultural difference indicated a less static and more process-oriented understanding of culture.[51] At the same time, the policy initiatives of the government moved from an overweening interest in maintaining cultural heritage (primarily focused on French Canadians) to a far stronger commitment to improving what it termed "race relations." Government funding for multicultural projects nearly tripled within a decade, with a far greater proportion of federal money allocated to programs dedicated to the betterment of relations between the "races."[52]

Alongside the increase and shift in government funding, there were concrete moves toward the entrenchment of multiculturalism at the constitutional and statutory levels of government. The 1982 Canadian Charter of Rights and Freedoms included two provisions related to multiculturalism. One of these, Section 27, explicitly linked the interpretation of the charter as consistent with the "preservation and enhancement of the multicultural heritage of Canadians." The Canadian Multiculturalism Act of 1988 even more directly affirmed the cultural diversity of Canada and the role of the government in "bringing about equal access for all Canadians in the economic, social, cultural and political realms."[53]

With a commitment to multiculturalism consecrated in the Charter of 1982 and securely established in the nation's statutes with the Multiculturalism Act of 1988, the first steps in building a new "global" Canadian order were taken.[54] The language used in reference to the new Multiculturalism Act explicitly linked multiculturalism with both national identity and national development. The connection of a new Canada with a new world order that involved international cooperation and increased economic prospects was similarly categorical. For example, David Crombie, the secretary of state of Canada and the minister responsible for multiculturalism, wrote in 1987:

> Dear fellow Canadians: I am pleased to introduce a Bill which, upon passage, will become the world's first national Multicultural Act. It contains the government's new policy respecting multiculturalism, an essential component of our Canadian identity.... Its intention is to strengthen our unity, reinforce our identity, improve our economic prospects and give recognition to historical and contemporary realities.... Multiculturalism has long been fundamental to the Canadian approach to nation-building.... Canadians are coming to realize that substantial social, economic and cultural benefits will flow from a strengthened commitment to multiculturalism.[55]

Crombie's words echoed the sentiments of Prime Minister Brian Mulroney, a strongly neoliberal politician who had emphasized the potential economic benefits of multiculturalism in 1986 at a conference called, appropriately, "Multiculturalism Means Business." In his speech, Mulroney was unequivocal about the pragmatic reasons for promoting a new, patently more strategic multiculturalism. He made the link between Canada's need for export markets and increased trading opportunities, with a more nurturing, interventionist government stance vis-à-vis the nation's ethnic members who might perhaps have links to "other" parts of the globe. The changing patterns of Canadian immigration made it more than likely that the "other" parts of the globe to which Mulroney referred would be located in Asia. The unambiguous gamble for increased business opportunities with the booming Pacific Rim countries through the ties of "multicultural" Chinese Canadians was couched in the language of the requirements of a newly deterritorializing nation-state and neoliberal entrepreneurialism:

> We, as a nation, need to grasp the opportunity afforded to us by our multicultural identity, to cement our prosperity with trade and investment links the world over and with a renewed entrepreneurial spirit at home.... In a competitive world, we all know that technology, pro-

ductivity, quality, marketing, and price determine export success. But our multicultural nature gives us an edge in selling to that world. . . . Canadians who have cultural links to other parts of the globe, who have business contacts elsewhere are of the utmost importance to our trade and investment strategy.[56]

Several government publications also connected the government's promotion of better race relations and the increasing immigration of wealthy Asian capitalists. In a statement by the Economic Council of Canada in 1991, *New Faces in the Crowd*, the authors juxtaposed three sets of statistics: the tremendous growth of immigration from Asia; the (positive) economic impact of the immigrant investor category for the Canadian economy; and increases in government funding for multicultural programs engaged with group acceptance and improved race relations.[57] The importance of the interconnections was implicit but clear. The deterritorialized state was embarking on a new ideological strategy that involved the mitigation of racial tensions *across* national borders, such as those that had surfaced around the movement of wealthy transnational Chinese immigrants and Asian capital into and out of Vancouver society.

Two new bills and a national campaign in 1989 and 1990 were also directly engaged with defusing racial animosity and educating people about racial discrimination.[58] The links between the new multiculturalist policy aimed at reducing racism and the growing friction in Vancouver surrounding Hong Kong Chinese investment were transparent. The national public education campaign to mark the International Day for the Elimination of Racial Discrimination, for example, was organized less than a month after two articles appeared in the *Globe and Mail*, a Toronto newspaper with a large national circulation; excerpts follow:

What, for Vancouver, is tomorrow? The answer: to call what is taking shape here startling is an understatement. Vancouver, barely past its 100th birthday, is going to become an Asian city.[59]

The Hong Kong immigrants are of a different breed from the usual new arrivals: they're rich. . . . Choice blocks of condominiums are being built that are sold only to Hong Kong buyers. Old houses are being bulldozed and replaced by unattractive megahouses for Hong Kong buyers. Hong Kong investment—about $800 million a year in the province, most of it in Vancouver—is gobbling real estate.[60]

In addition to a flourishing of proimmigrant, antiracist government organizations in Vancouver in 1989 and 1990, several private institutions began to promote antiracist global multiculturalism in the city as well. The desire to facilitate capital accumulation and circulation by reducing racial friction was evident in the educational pamphlets and commissioned reports of these organizations. So too was a clear effort to aid the city's integration into the global economy. Vancouver's Laurier Institute was the most prominent and well financed of these groups.

The Laurier Institute came into legal existence in mid 1989, but according to the executive director it had by then been operational for a year. Its goals appear in a number of brochures and publications; for example, from a 1990 newsletter: "To contribute to the effective integration of the many diverse cultural groups within Canadian society into our political, social and economic life by educating Canadians of the positive features of diversity." Orest Kruhlak, the executive director of the institute, specifically mentioned in our 1991 Vancouver interview the attempt to defuse potential racial friction as a primary goal of the organization. He cited the potential social problems caused by increased racial diversity, and the hope of eradicating these problems with an extensive public education campaign: "Nobody seemed to be looking at the long-term implications of increased diversity. What we wanted to do was say how can we start working with some of the issues that might come forward in the future with the idea of trying to get ahead on the issue and do research and educational programming to try and prevent the problems that we have come to understand were going to be major problems in the future."

One of the first projects commissioned by the institute was a study of real-estate price increases in Vancouver. According to Kruhlak, this study had not been planned, but was in response to "a growing and emerging problem." In the report, "Population and Housing in Metropolitan Vancouver: Changing Patterns of Demographics and Demand," the author's results indicated that the rising house prices were the result of demand from the aging Canadian postwar baby boomers. The author, David Baxter, wrote in the executive summary: "Regardless of the level of migration assumed (none, normal, or high) and regardless of the level of household headship rates assumed (constant or increasing) it is the demographic process of the aging of the post war baby boom into the 35 to 44 age group (1986 to 1996) and then into the 45 to 54 age group (1991 to 2006) that will determine the characteristics of changes in housing demand in metropolitan Vancouver in the future."[61]

Although this paragraph is quite general, indicating that demographic change was responsible for changes in housing demand in Vancouver (with the implication of increased house prices), the following paragraph in the summary demonstrated the persuasive rhetorical strategy at the heart of the study. Here, Baxter made it clear that the report was concerned with showing not so much the possible reasons for price increases as what were *not* possible reasons. It was unacceptable in the dictates of global multiculturalism and Canadian national narratives of social tolerance to identify or pinpoint a particular group. In an avuncular, warning tone, Baxter wrote: "If we seek someone to blame for this increase in demand, we will find only that the responsible group is everyone, not some unusual or exotic group of residents or migrants. In fact, there is no one to blame: the future growth in housing demand is a logical and normal extension of trends in the nation's population."[62]

The effect on the media of Baxter's warning statement was immediate and immediately conveyed to the public. Nearly all the major Canadian newspapers commented on the findings of the report, part of a series of research projects jointly sponsored by the Laurier Institute, the Canadian Real Estate Research Bureau, and the Bureau of Applied Research (in the Faculty of Commerce and Business Administration at the University of British Columbia). The *Vancouver Sun* and the *Globe and Mail*, still recovering from allegations of racism in an earlier series written about Hong Kong, instantly connected Baxter's statement about "unusual or exotic groups" with wealthy immigrants from Hong Kong, a connection not explicitly made anywhere in the report. The *Globe and Mail* wrote in November 1989: "Aging baby boomers, not foreign immigrants, are the main reason Vancouver housing prices are rising, a study says.... [The study] was prompted by public complaints that home-buying by affluent Hong Kong immigrants had been forcing up Vancouver home prices."[63]

The other Laurier reports, all of which focused on housing and real estate in Vancouver, were also broadcast nationally in several forums. The report by Professor Hamilton, "Residential Market Behavior: Turnover Rates and Holding Periods," claimed that immigration levels did not contribute to speculation in the housing market. *Vancouver Sun* headlines declared soon after, on April 19, 1990, "Foreign Buyers Absolved," and the local Vancouver paper, the *Courier*, proclaimed on April 22, "Home Speculation Not Immigrants' Fault."

A study by Dr. Enid Slack, the fourth in the Laurier series, showed that the levies on developers (for financing water supply systems, sewage treatment plants, etc.) were often passed on to new home buyers in increased house prices. The *Real Estate Weekly* in Vancouver wrote of these findings: "Slack's report is part of a major study commissioned by the Laurier Institute to determine whether any basis exists for suggestions that Chinese immigrant buyers are driving up Vancouver real estate prices. So far, the Institute has found, 'in fact there is no one to blame; . . . the responsible group is everyone.'"[64]

The reiteration of Baxter's statement that no one was to blame operated like a mantra for warding off the evil spirit of racism. But the statement in fact operated on a number of levels. Denying that blame for higher real-estate prices could be pinned on anyone in particular achieved two results: first, those who disagreed were not joining the valiant effort to defeat racism and uphold multiculturalism, and thus could be seen as potentially racist and unpatriotic; and second, since everyone was to be held responsible and no one to blame, there was no obligation and no need to unmask the agents or systems involved in the process that had, in fact, led to higher prices. The workings of capitalism thus remained opaque, the agents involved in capital transfer remained faceless, and the spatial barriers and frictions that might have disrupted neoliberal urban policy and the free flow of capital over and through municipal and international borders were eradicated.

The implications of Slack's report went one step further. Not only was no one to blame for the unfortunate (for house buyers) rise in house prices, but if anyone should be held accountable, it was city government. Although house prices doubled and tripled in a single year, adding hundreds of thousands of dollars to prices in certain neighborhoods, Slack's findings focused on the development costs imposed by *municipalities* on new housing projects. These costs ranged from approximately C$1,500 for new homes in Burnaby, to about C$12,000 for a new home in Richmond. Furthermore, the costs were not imposed in many communities in Vancouver or North Vancouver. Nevertheless, the March 23, 1990, *Vancouver Sun* commented on Slack's report under the headline "Study Finds Extra Charges Placing Heavy Burden on New Home Buyers." Here, the state-imposed "extra charges" were blamed for the rising house prices; an overly controlling and interfering municipal government was positioned as the primary source of Vancouver's housing woes.

The Laurier reports are indicative of a virulent form of neoliberal urban discourse in Vancouver in the late 1980s and 1990s. In addition to the rhetoric that labeled any form of state intervention problematic, as causing unnecessary blockages and costs by disrupting market forces, these reports portrayed the city as rife with entrenched anti-immigrant and racist superstitions. The struggles over housing prices, affordability, livability, and quality of life—the literal spaces of the city—were inescapably racialized. Further, this process of racialization was so irrational and so deeply sedimented in the landscape, it could finally be vanquished only by eradicating spatial sentiment altogether. Thus neoliberal discourse, as represented in the Laurier reports, evoked the spacelessness of neoliberal theory as the utopia of a "raceless" society where the market alone would arbitrate price, and who could live where.

Why would the Laurier Institute, whose express mission and role was to "promote cultural harmony in Canada" and "encourage understanding among and between people of various cultures," commission these reports? Although my sources indicated that the Laurier Institute was founded by "a group of businessmen," the organization's brochures did not identify the founders by name. The list of the board of directors in 1990, however, included thirteen people and their positions, among them four lawyers in major law firms, three executives in large corporations, two investment and management counselors, and one real-estate executive. Of the corporations or firms represented, nine were directly or indirectly involved with Hong Kong business or investment. Among the seven founding donors (C$25,000 or more), there was a similar overlap with Hong Kong business concerns.[65]

The Laurier Institute persuaded people of the benefits of neoliberal multiculturalism through a number of media, including a video and curriculum guide for use in the schools, "Growing Up Asian and Native Canadian."[66] The dissemination of general information was made public and broadcast via the media. In addition, the information generated from commissioned research was available to companies who became corporate members of the Laurier Institute. One brochure, "Cultural Harmony through Research, Communication, and Education," listed under "Corporate Member Benefits" the economic advantages of having insider information on Canada's cultural diversity. As in Mulroney's speech quoted earlier, the emphasis on the global economic potential of contributing to the multicultural ideology was clear: "The cultural diversity of Canada's population has brought, and continues to bring, significant change to the Canadian workforce and the Canadian marketplace. Companies which recognize the potential of this diversity and act accordingly will have enormous advantage over those who do not. Membership in The Laurier Institute offers assistance in terms of both recognizing the potential and implementing programs which will deliver that advantage."

Other corporate members and major supporters of the Laurier Institute included the Bank of Nova Scotia, the Canadian Maple Leaf Fund, Ltd., Concord Pacific Developments, Grand Adex, Hong Kong Bank of Canada, Pacific Canadian Investment, and the Royal Bank of Canada. Those who helped fund the reports on housing prices and municipal levies included the major real-estate foundations in Vancouver. All these corporations had and continue to have major stakes in Hong Kong and in the ongoing flow of people and capital from Hong Kong into and through Canada.

The Laurier Institute derided state "intervention" in the taxation of housing development, the primary mode through which the municipal government was able to garner funds for social services. At the same time, however, it drew from and extended a liberal philosophy of antiracist multiculturalism as part of a strategy to aid global integration. By reducing the anger over the perception that Hong Kong Chinese immigrants were buying real estate for speculative purposes, the institute sought to reduce racial friction that impeded the continued investment of foreign capital in the region. Invoking the benefits of tolerance and patience, disseminating blame for unpleasant shifts in the architecture of landscape and workplace, and promoting better "race relations" operated in this case to camouflage the workings of capitalism and of a newly scaled allegiance to the global economy. Cruz writes of these types of activities: "What I propose is that we step back and look at multiculturalism as part of a *social logic* of late capitalism and as a cultural feature at the intersection of economic globalization and the fiscal-domestic crisis of the state."[67]

## INSTITUTIONS OF COSMOPOLITAN CITIZENSHIP FORMATION

Foreigners in large numbers are in our midst. More are coming. How are we to make them into good Canadian citizens? First of all, they must in some way be unified. Language, nationality, race, temperament, training, are all dividing walls that must be broken down.
    —James Woodsworth, *Strangers within Our Gates*[68]

Cross-cultural training programmes have been developed to inculcate sensitivity, basic savoir faire, and perhaps an appreciation of those other cultures which are of special strategic importance to one's goals.
    —Ulf Hannerz, "Cosmopolitans and Locals"[69]

If the Laurier Institute focused on producing a neoliberal multiculturalism both to guide Canadians toward greater tolerance of wealthy Chinese immigrants and to strategically use diversity for a global agenda, the Meet with Success program in Hong Kong emphasized the guidance of the immigrants themselves. Before leaving Hong Kong, emigrants bound for Canada were schooled in the appropriate Canadian way of life. Acculturation was promoted through a series of lectures, videos, and pamphlets that emphasized the essential character of Canadianness and the necessity for immigrants to understand and assimilate that character. With avuncular authority, the future immigrants were reminded of the many obligations that accompanied the freedoms they would experience living in Canada. Adaptation to a British Canadian way of life, and to the values and norms of a secular liberal society, was of primary importance.

Meet with Success was produced by the Canadian Club of Hong Kong in the spring of 1990. Every Hong Kong applicant to Canada who was accepted as an immigrant was invited to attend a Meet with Success program. The invitation was extended by the Commission for Canada in the same packet as the Canadian visa. Attendance was voluntary, but there was a 90 percent turnout; as a program volunteer confided to me in a 1991 interview in Hong Kong: "Some think it is required because it comes from the Commission." The program was offered every week and consisted of a video and slide show and a question-and-answer session. Upon arriving at the meeting, each person received an informational packet with a long survey entitled "Meet with Success: Personal Inventory of Values."

When I attended the program in Hong Kong, I realized that I had already seen the video, which is shown weekly at immigration and investment seminars in Vancouver.[70] The video, *Being Canadian*, depicts a young Hong Kong couple arriving in Vancouver. The moderator, a recent immigrant to Vancouver from Hong Kong, says, "Let me tell you about Vancouver." He is filmed in his home, an average-looking house on the east side of the city. He says reflectively, "I love this city or else I would not have moved here, but lifestyles here are different."[71] The implication is that he chose to move to Vancouver for *love* rather than for the pursuit of profit. Nonetheless, despite his love for the city, adaptation has been difficult. For overcoming these difficulties, he has been rewarded with a house on the east side and is shown as completely content.

Other images of Vancouver are used in a similarly paternalistic and didactic manner. There is a scene of shopping at the vegetable markets in Granville Island, while the narrator intones: "People don't bargain loudly. They are polite to each other." In another scene, several cars are shown at an intersection; the narrator observes: "They are patient. Each waits his turn so it doesn't become a traffic jam." There is a scene in a garden, where a white woman educates a Chinese man about gardening and Christmas. The narrator says thoughtfully, "She not only helps her neighbors, she also makes her garden beautiful." Most interestingly, there are several images of flowers, large trees, and natural landscaping, obviously filmed in west-side Vancouver neighborhoods. The narrator says, "This is a Vancouver neighborhood." While the camera pans over the gardens and homes to an image of a "monster" house, the narrator says: "Everybody likes to live in a nice place. Newer, bigger houses are a new trend. Sometimes the new houses look out of place."

According to two of my sources who were involved with planning Meet with Success in 1989, the program was designed as a direct result of the tremendous controversy and negative international publicity that flared up in Vancouver in 1988 and 1989 over the so-called monster houses.[72] One woman associated with the program said in our July 1991 interview in Hong Kong that in addition to problems surrounding the monster-house issue, there were numerous other social conflicts, including "the ostentatious flaunting of wealth" and "double parking by Chinese students outside schools who just paid the fines and continued parking there." She noted that it was "not a smooth adaptation" when Hong Kong Chinese immigrants arrived in Vancouver, and in fact, there was great resentment by many Canadians of the "yacht people." According to her, the Canadian Club organizers in Hong Kong were concerned about this friction, especially as they did not want Canada tarred with the same racist image that was being applied to Australia at that time.

Another organizer told me that helping Hong Kong immigrants learn about Canada was an attempt to avoid confrontation and "make life easier" for all involved. She noted that various visual images of Canadian life demonstrated cultural norms and unacceptable violations of those norms (such as removing trees and building large houses) without verbalizing the problem, which might be interpreted as racist. The images of unacceptable violations were decided on by members of the board; according to a white Canadian member, the Chinese members of the board wanted to be tougher on cultural "problem areas" such as spitting. In the final decision, however, this stance was rejected as too patronizing.

This program tried to reduce potential frictions involving cultural differences by dispensing advice to the immigrants before they left Hong Kong. Legal acceptance into Canadian society was insufficient; the new immigrants had to become *culturally* Canadian to avoid disrupting the urban fabric and the social coherence necessary for continued international investment. Exclusion from the national community on the basis of racism, rather than cultural difference, was a prohibited subject for public discussion. The belief in *cultural* citizenship relied on an assumption that cultural differences and frictions were superficial and transformable in relation to a profound and fundamentally neutral liberalism. The Canadian way of life was represented as value free and assumable without the loss of a Chinese way of life; further, it was presumed that this assimilation would result in mutually advantageous social harmony for everyone. Thus the "good" (neutral and universal) values of hard work, education, and family could be retained, but the "bad" (particularist and tribalist) characteristics of rudeness, impatience at traffic lights, and a dislike of gardening had to be discarded.

The Canadian cultural norms of a socially liberal, tolerant, humane, and ultimately *pastoral* society were taken for granted. As these values were implicitly equated with a civilized and socially cohesive national community, they were normalized through time and hence became invisible. Liberal multiculturalism in this case consisted of an imagined urban space that was assumed to be the only acceptable territory for those who desired inclusion in the political unity.[73] Those who engaged in unfamiliar spatial actions or relations would be denied access to the universal polity until they were properly educated into the appropriate forms of spatial distinction.

In addition to the video, the "Personal Inventory of Values" survey directly addressed concerns about lifestyle and consumption patterns. Written in English and Chinese, it represented "a checklist for your reflection before and after you arrive in Canada." The respondent was asked to read several statements and circle "Important" or "Not So Important" in the margin to indicate his or her "personal values." The didactic intent of the survey was clear: "These are personal values; the choices are yours but please make the commitment to understanding the economic, social and political environment within which you will be making them. Detailed information is contained in the accompanying 'Fact Sheets' in this kit."

One of my informants involved with the program said in our July 1991 interview: "It is their choice as to whether or not they want to modify their behavior. We are just alerting them as to possible conflict areas." The questions in the survey ranged from the work environment, housing, and family life to multiculturalism, social skills, and the assimilation process. Although the questions could be circled "Not So Important," the wording made it obvious that this answer was not acceptable. The immigrant's responsibility for understanding the underlying message of acculturation was explicit. That this strategy would be economically and socially beneficial for all concerned was also a clear message of the survey.

The "right" answers to the "Personal Inventory of Values" were located in the fact sheets. For example, in the category "Housing Options," statements such as the following educated the immigrants about the appropriate feelings to have concerning their homes: "Houses have historically been a good investment in Canada but the gains are not all financial. Communities of house owners tend to be stable and relatively friendly. Relationships with neighbours are important. Houses are a big commitment in time and lifestyle as well as money." The warning against the disruptions caused by speculation was clear. As in the video, the correct reason to purchase a house was for love of the neighborhood or the house itself, not for financial gain. In the section "Choosing a House and Neighborhood," the concern about frictions caused by the monster houses was broached in an even more didactic and warning tone. In this case, if the norms of the society were not willingly assumed, the legal system would be brought to bear on the most egregious offenders of good taste: "The style of the houses and the amount of green space are serious considerations. Conformity to existing standards is a custom that in some cases is prescribed by municipal by-law. Plans to significantly enlarge houses, change the style of the exterior decor or cut down trees may be offensive if it is legal at all."

In the "Multiculturalism Fact Sheet," the responsibility of the immigrants to become acculturated into a British Canadian secular society was underscored at the same time that the advantages of a strategic Canadian cultural pluralism were upheld. Immigrant responsibilities included learning the language and history of the host society, participating in community and political life, contributing to social security through taxes and community involvement, and tolerating cultural, racial, and religious differences. In the last section, "Ways of Preserving Your Own Cultural Background," the ways in which both the responsibilities of acculturation and the rights of multiculturalism could be accommodated were enumerated: "Join Chinese community groups. Teach your children the Chinese language. Participate in multicultural festivals such as Toronto's 'Caravan.' Listen to Chinese radio; watch Chinese TV; and read Chinese newspapers."

The message of assimilating the cultural values of the host society was fairly straightforward; it was possible, according to Meet with Success, to be both Canadian and Chinese through a strategic manipulation of cultural citizenship and the meaning of multiculturalism. Among the many silences in the video and packet was the question of the continuing transnational movement of people and capital between Hong Kong and Vancouver. Racism and class conflict relating to this ongoing movement were purposely disregarded as private issues separate from the cultural differences profiled in the seminar. This move to conflate cultural and racial conflict and to separate cultural and economic concerns has a long history in Canadian multicultural policies.[74]

The fashioning of the middle-class Hong Kong immigrant was promoted by the wealthy patrons and sponsors of both Canadian and Hong Kong society.[75] The video *Being Canadian* was filmed in Vancouver, written by a member of the Canadian Club in Hong Kong, and funded by the Asia Pacific Foundation, an organization involved with Canadian–Pacific Rim connections and listed as a corporate member and major supporter of the Laurier Institute. Patrons of the Meet with Success program (contributing C$25,000) included ten people or groups, many of whom also appeared on the Laurier Institute funding list. Major contributors included the Canadian Maple Leaf Fund, Ltd., Michael Goldberg, twenty-one people connected with the Canadian government, and three people associated with the Canadian Imperial Bank of Commerce.

In the Meet with Success program it is possible to discern the articulation between state efforts at a new form of deterritorialized subjectivity formation and the extension of neoliberal capitalist hegemony. The desired Canadian subject is one who exists in a nonfrictional space; this is the type of space most conducive to the self-perceived level playing field of neoliberalism. The contemporary rights to (Canadian) citizenship and welcome into the political community were proffered to those willing to leave behind a cultural allegiance that could be connected to the particularisms of a specific place. Canadian citizenship and Canadian liberalism were equated with a strategic cosmopolitanism that was both spaceless and value free. This is the fantasy of global neoliberalism without the drag of local particularisms. It represents a strategic multiculturalism premised on the "neutral" values of neoliberalism—in other words, a disembodied political belonging in a global void.

## THE LIMITS OF LIBERAL MULTICULTURALISM

Liberal multiculturalism in Canada has evolved through time, beginning with its foundational presumption of a recognition of difference within the public sphere. Recognition of identity was based on singularity, on the conceptual framework of individuals and groups operating as discrete, clearly differentiated rational units. This process of differentiation created singularities that the state could manage rationally within the context of the national public sphere, a process of domestication and control which served to constantly reify the territorial state and its boundaries.[76]

Liberal multiculturalism, in this context, was part of a larger project of state formation in Canada ongoing for over a century. The fragmented nation cohered through the narratives of tolerance and fairness, and the underlying commitment to the core political values of Britishness. In the current moment of transformation caused by the transnational movement of immigrants and capital, and by new kinds of neoliberal state pressures and responses, the fundamental contradictions within liberalism—including the unresolvable differences between a civic understanding of individual rights, a communitarian conceptualization of collective rights and the community good, and a nonindividualist understanding of social roles and networked relations across space—came to the fore and could no longer be dissimulated.[77]

The idea of a liberal national community in Canada was built upon the foundation of harmonious relations extended toward differing "racial" groups (such as the French), and of the endlessly expandable, ungrudgingly generous inclusion of those groups within the tenets of liberal democracy. The profound British-centrism of this ideal "multicultural" national community, however, became increasingly evident in the rancorous debates surrounding the proposed Meech Lake Accord. The allowance of "difference" within the accord for the purposes of collective cultural survival for the Québécois was antithetical to the model of procedural liberalism most familiar to English Canadians, which was recently adopted in the Canadian Charter.[78] As a result, there was a struggle over what should be given precedence: the "distinct society" clause of the accord, or the basic proceduralism of the charter. As noted by the Canadian political philosopher Charles Taylor:

> The resistance to the "distinct society" that called for precedence to be given to the Charter came in part from a spreading procedural outlook in English Canada. From this point of view, attributing the goal of promoting Québec's distinct society to a government is to acknowledge a collective goal, and this move had to be neutralized by being subordinated to the existing Charter. From the standpoint of Québec, this attempt to impose a procedural model of liberalism not only would deprive the distinct society clause of some of its force as a rule of interpretation, but bespoke a rejection of the model of liberalism on which this society was founded. Each society misperceived the other throughout the Meech Lake debate. But here both perceived each other accurately—and didn't like what they saw. The rest of Canada saw that the distinct society clause legitimated collective goals. And Québec saw that the move to give the Charter precedence imposed a form of liberal society that was alien to it, and to which Québec could never accommodate itself without surrendering its identity.[79]

In the Meech Lake debates and the ensuing negative vote (the accord was never legislated), the British orientation of the nation, as well as the limits of multiculturalism (as circumscribed within a liberal-proceduralist framework), became immediately apparent. In the course of the Meech Lake discussion, differing strands of liberalism quickly came into conflict: the first privileged a neutral, individual, rights-based, and proceduralist vision stemming from the British tradition; the second foregrounded a communitarian, group-based vision which allowed for a concept of the "good society" and the right to group separation to ensure cultural survival over time.

Despite the fundamental national conflict manifested in the Meech Lake debates, however, the "proceduralist" variant of the liberal philosophy of multiculturalism maintained its position as the narrative of national unity and the predominant governmental rhetoric and policy in Canada. Support for a proceduralist variant of multiculturalism was included in a section of the Canadian Constitution Act of 1982, and in the Multiculturalism Act of 1988. In this statute and in numerous other government texts and statements, the earlier seeds of nation-state formation and of the ongoing attempt to reconstitute a "beloved" national community remained firmly embedded in the multicultural language. Multiculturalism in Canada through the mid 1980s thus remained doubly inscribed: it was inherently nationalist in purpose and orientation, and it was also based on a proceduralist model of the rights-bearing individual in liberalism that privileged British philosophy and culture as the national norm.

These particular liberal assumptions were challenged, however, by the arrival of the contemporary Hong Kong Chinese immigrants. For many of these immigrants, concepts of personal identity were conditioned by the importance of social roles and extended community relationships, as much as by conceptualizations of the rights-bearing individual.[80] Further, in transnational societies such as the Hong Kong Chinese "astronaut" group, these social roles can play out through vast networks extended across space, including across multiple international borders.[81] The social norms associated with a sense of personhood or individual identity are thus not focused in precisely the same way on individual rights as in the West, nor are they inherently implicated in the performative construction of the liberal nation.

How can multiculturalism, within the framework of Canadian social liberalism, contain and manage a set of transnational relationships, both material and discursive, when the logic of its formulation is that of public recognition of the rights-bearing individual within the territorial container of the nation-state? It cannot. I believe that liberal multiculturalism in Canada exists in profound conflict with the idea of the person and the community as multidimensional and multispatial, as a set of linked relationships and allegiances that extend over space. Canadian liberal multiculturalism is premised on an understanding of the public sphere as the discrete space of the nation-state; this premise cannot hold alongside a conceptualization of the public sphere as the networked space of transnational imaginings.

Transnationalism is about multiplicities, about the multidimensional types of social and economic relations that work across scales. Transnational identities do not necessarily conform to a singular or contained entity, nor do transnational formations belong to a contiguous, uniform, or contained space. These shifts in scale challenge liberal projects of domestication by identity-based unit, and also the processes through which people are *made* individual (through state technologies of counting, healing, paying, seeing, etc.). With transnational flows into any given society, such as Canada's, different forms of relations begin to coexist and metamorphize; they don't supplant former understandings, but rather draw attention to the normative assumptions which underpin them. It is the new spatial movements and relationships across, between, and within multidimensional scales that threaten and inevitably transform the static notions of space, person, and nation implicit within liberal multiculturalism.

Liberal multiculturalism is a fundamentally additive political philosophy; differences can be "added into" the great national stewpot, and social liberals will generously tolerate these additions and be sensitive to the special needs of Others. But more than this, as scholars like Mehta have shown, liberalism has evolved as a nationalist philosophy with imperial ambitions, and each addition to the multicultural pot forms another layer of state formation in both space and time.[82] Disruptions to these secret secretions occur through the failure of bodies and entities to join the (imperial) project, and this failure occurs spatially as well as temporally, although it is the temporal dimension that is most often debated by both liberals and their critics.

Because most liberals do not adequately engage with the issue of space, the contemporary contradictions within social liberalism that are exacerbated by globalization, neoliberalization, and transnationalization remain undertheorized. In Taylor's reasoning, for example, it is those cultures that extend "over some considerable stretch of time" that do or do not deserve "equal respect" within a politics of multicultural recognition.[83] This judgment, however, generally begins in a spaceless void. Thus Taylor makes the mistake of equating a *socially liberal* multiculturalism, which rests on the imagined community of the nation and on the subjectivity formation of a multicultural self, with a *neo*liberal multiculturalism, which exists in a vacuum. *Neo*liberal multiculturalism seeks to erase the complexities of history and geography and start as if from the same originating place and time, where those people and cultures "that have provided the horizon of meaning for large numbers of human beings ... over a long period of time" can be perceived and judged as equal.[84] As Bhabha notes in a critique of Taylor, this starting point of a "historically congruent space" and "level playing field" ignores the central importance of the disruptive cultural voices and nonmajoritarian cultural interventions that constantly impede and alter a normative sense of cultural and national time.

> At the point at which liberal discourse attempts to normalise cultural respect into the recognition of *equal cultural worth*, it does not recognise the disjunctive, "borderline" temporalities of partial, minority cultures. The sharing of equality is genuinely intended, but only so long as we start from a historically congruent space; the recognition of difference is genuinely felt, but on terms that do not represent the historical genealogies, often postcolonial, that constitute the partial cultures of the minority.[85]

Bhabha's critique of Taylor rests on his familiar heralding of the Bakhtinian-style disruptions of minority or marginalized "partial cultures" that are "out of time" because they are "in between" the temporal rhythms established for the nation-state.[86] Unlike Taylor, who is primarily interested in stability and survival—the cultural survival of minority cultures (such as the Québécois), and the stability of the nation (Canada)—Bhabha is centrally concerned with instability, differentiation, and transmogrification. Yet Bhabha's critique of the underlying homogeneous temporalities of Taylor's line of reasoning is limited by its emphasis on cultural time. He misses the important axis of space, and the multiple ways in which partial, minority cultures are not just "out of time," but also "out of place" vis-à-vis national imaginings. For Bhabha, space is not productive of time but only the reverse. He thus has little regard for the production of space or the sedimentation of history in space, and his sense of the "hybridising potential" of the dialogic is hence deprived of its more exciting possibilities.[87]

## CONCLUSION: TOWARD A NEOLIBERAL MULTICULTURALISM

One of my main interests in multiculturalism, in this chapter, is in the ways that, as a concept, it has been put into the service of the liberal state. What ideological work does the concept do, and why is it currently being transformed in the context of globalization, transnational migration, deterritorialization, and other kinds of neoliberal pressures? Multiculturalism in Canada has functioned through time as a key national narrative of coherence and unification. State legitimacy was increased through the ability of politicians and bureaucrats to unify internal differences and direct them into a single project, that of state formation. State actors and institutions engaged in this project through the regulation of individual and (carefully delineated) group rights, as evidenced in the philosophy, practice, and legislation of tenets such as multiculturalism.

For Canada, multiculturalism was an official state doctrine which allowed an uneasy truce to be formed between the original two colonizing powers, the British and the French. It also represented an effort to inculcate immigrants into a national mosaic, wherein difference was professed as welcome and even advantageous to state development. The concept of multicultural citizenship served as an example of the tolerant and munificent liberal state, ever willing to open its doors to outsiders, and to accept and protect cultural difference. It was part of a broader narrative of liberalism and the freedom of choice for the individual, and through this narrative served to perform the liberal state and to create the sense of a unified, tolerant, and coherent nation, despite the multiple differences evident among its citizens.

Multiculturalism also operated as a fundamental institutional and conceptual tool that gave the state enhanced ability to *control* difference.[88] As a conceptual apparatus, it allowed the state to set the terms of the "difference debate." These terms remained highly individualistic, concerned with individual rights and preferences, the right to choose and display difference with respect to individual identity. Cultural pluralism was encouraged, but only so long as the included groups followed certain rules and were willing to be contained by the strict parameters of liberalism—that is, to "accept" liberalism as a fundamental philosophical starting point.[89] And while group difference was acceptable for "cultural survival" (e.g., in the case of the Québécois), it was acceptable only in certain carefully circumscribed times and spaces (e.g., within the province of Québec).

Canadian multiculturalism through the early 1980s involved constituting a subject who was tolerant of difference, but a difference framed within national parameters and controlled by the institutions of the state. The multicultural subject of this time believed that cultural pluralism was good, or at least necessary, for national development, and was able to work with others to find sites of commonality, despite differences. These understandings had a certain logic when implemented within the format of the territorial state. But what happens when "the community" is no longer necessarily a national one, and the interest of state actors in disciplining populations and regulating the actions and relations between individuals shifts to a larger scale?

My argument in this chapter is that liberal multiculturalism grew out of a logic of interprovincial fragmentation and pluralism, and was effective in the service of both state formation and economic development through the early 1980s. Multiculturalism operated effectively as an instrument of state formation on a number of levels: as a national narrative of coherence in the face of British-French and then immigrant "difference," as a broad technology of state control of difference, and as one of many capillaries of disciplinary power/knowledge concerning the formation of the state subject. In all of this, but especially in the constitution of national citizens able and willing to work *through* difference *for* the nation, the socially liberal philosophy and practice of multiculturalism was a strategic partner in the growth and expansion of a Fordist capitalist regime of accumulation.

However, with the rise of transnational lives, deterritorialized states, and neoliberal pressures in the past two decades, this type of state subject has become increasingly irrelevant. The particular form of what I have termed "liberal multiculturalism"—one jointly bound up in the constitution of the nation, the tolerant national self, and the formation of a regime of accumulation regulated by the state—is evolving into something qualitatively different. Liberal multiculturalism, a spatially specific ethos of tolerance contingent on the history and geography of a city and a nation, is now rapidly morphing into *neo*liberal multiculturalism, the "progressive process of planetary integration."[90] It reflects a logic of pluralism on a global scale, and a strategic, outward-looking cosmopolitanism.

The transformation of multiculturalism became visible in Vancouver as a result of the struggles over spatial practices in the city with the arrival of the unfamiliar bodies and beliefs of Hong Kong transnationals. State politicians, business leaders, and older residents intervened in these struggles in a manner that evoked the shifting conceptualization and rhetoric of plural tolerance. Rather than the socially liberal principle of getting along for the sake of national unity, the new rationale for state multiculturalism was maintaining the unrestricted flow of global capital into and out of the landscape. Racism and unacceptable (friction-producing) cultural difference were regulated and disciplined, not for the purposes of capital accumulation in the development of the state, as in the Fordist moment, but rather to keep the city and the state safely centered as key nodes in the expanding networks of global capitalism.

# 4    Disturbing the Liberal Territory of Land Governance

URBAN LAND GOVERNANCE is a key process for state hegemony, because it is through the institutions and practices of urban planning that the state is able to produce both profit and legitimacy. Municipal politicians and planners, working within the constraints of urban, provincial, and federal mandates, retain zoning rights, building rights, and the general right to manage and regulate urban consumption processes in the city. Yet in a liberal democracy these forms of management can be implemented only through the constant manipulation of urban consensus. With the immigration and rapid capital influx from Hong Kong in the late 1980s, the usual configurations of consensus on the "rationality" of various kinds of land use were altered in Vancouver by shifting class and neighborhood alliances. The disturbance of these alliances was integral to the reworking of spatial hegemony and the struggles over liberal and neoliberal policy in the city.

New alliances included those between some of the recent Hong Kong immigrants, neoliberal politicians and planners, and various members of the real-estate industry. At the same time, there was also considerable fracturing and reworking of established coalitions, such as those between west-side, upper-class interests and pro-growth political parties like the Non-Partisan Association (NPA). Since west-side neighborhoods were also experiencing unwanted growth, many wealthy homeowners in these areas felt betrayed by their pro-growth municipal representatives. Politicians were thus confronted with new kinds of disagreements over zoning, razing, building, regulating, tree removal, and other instruments of state control, in which the heretofore normalized processes of informal consensus building in the interests of a dominant class were no longer necessarily effective.

The hegemonic discourse of land governance in Vancouver in the 1970s had an aura of social liberalism, at least partly because of the multiscaled configurations of liberal political power present during the TEAM years. During this decade, the organization, dispensation, and planning of urban land—which has always involved a coalition of forces that included, most prominently, the real-estate industry and big business—appeared to tilt ever so slightly in the direction of social entitlements and economic redistribution. This was true especially in the realm of public housing provision and the discourse of a universal right to a stable and humane urban environment.[1] A strong narrative of regulation and oversight for the urban community, and by implication for the national community, ran through the pronouncements of municipal politicians and planners; alongside it ran the continuing mantra of growth, tolerance, and progress as interconnected and inherently positive. The development of Vancouver was linked with the successful development of Canadian society, perceived by many during these years as a "precarious experiment" in political democracy, as it navigated the turbulent waters of Québecois "difference" and increasingly lopsided power relations with the United States.[2] Progress, harmony, development, and rational state regulation and control were firmly articulated with the vision of a compassionate and livable city and a viable nation.

The socially liberal coalition of actors and institutions involved in urban land governance during these years, while still connected with the real-estate industry, maintained a high-profile image of active social welfarism. This image was the result of a rhetoric of universal urban entitlement and of the actual implementation of a number of partly successful and highly visible urban social projects.[3] Of interest to me in this chapter are the practices of social and economic redistribution and the liberal discourse of urban entitlement and membership, bound up with the narrative of Canadian national development. For by the mid 1980s, this socially liberal constellation of practices and ideas concerning the optimal social welfare of the city of Vancouver and of the Canadian nation came under increasing attack.

Neoliberal politicians, in conjunction with the real-estate industry and other urban actors, sought to roll back the formulations of a redistributionist, nationally oriented social liberalism, and to roll out a new agenda of state regulation and rational control on a global scale. This new agenda for urban land governance involved obliterating all barriers to the global circulation of capital and bodies in urban space, and creating an outward-oriented urban enterprise zone.[4] In the long-term neoliberal vision, urban governance itself would be unnecessary, as market forces would become self-regulating. However, in the immediate construction of this reputedly level and neutral space, the rational and regulative control of the state was imperative; this control required winning "consent" from the multiple players involved in urban public policy.

Thus in the decades of the 1980s and 1990s, a battle was born. It involved the future of land governance in Vancouver; the appropriate role of the state and of urban residents in issues of land governance; and the forms, symbols, practices, and narratives through which these roles should be articulated and comprehended. The ensuing hegemonic struggles were not abstract, as they can be at larger scales, but directly proximate and important to all involved. This proximity matters; it affects the forms that spatial struggles take and the ways in which hegemony is produced and contested through them. In the context of the transformation of houses, trees, and land in Vancouver, different groups interpreted and drew on differently scaled understandings of Canadian territory and community (of neighborhood, city, nation, and global diasporic networks); they used these interpretations to try to win advantage for their own urban agendas.

In this chapter I discuss three cases in which different actors and institutions manipulated and contested the discourse of urban land governance. In all three, the players drew on their own interpretations of Canadian liberalism to bolster their claims concerning the appropriate organization and control of the Vancouver landscape. In the first example, which involved a lengthy newspaper series in the *Vancouver Sun*, an alliance of developers, businesspeople, politicians, and academic planners promoted an odd philosophical brew of communitarianism and neoliberalism in their effort to counter what they represented as the excesses of *civil* liberalism, that is, the extreme expression of individual rights. While they supported state regulation to discipline excessive democratic liberalism, which they positioned as bad for the community, these growth advocates upheld the *economic* private property rights of freedom from state interference. Further, they drew on a convoluted mélange of communitarian narratives of the "good society" to promote a neoliberal emphasis vis-à-vis the certitude of Vancouver's integration into the global economy and the importance of behaving "rationally" to facilitate and profit from this inevitable outcome.

In the second case, wealthy residents of Shaughnessy Heights, an old, established west-side neighborhood, attempted to rezone their neighborhood to make further development unlikely. The property rights association connected with this neighborhood, along with the Vancouver Planning Department, convened a series of public hearings on the proposed downzoning to bring this issue to the public and to manifest the openness of the democratic process concerning land management in the city. At the hearings, however, numerous recently arrived Chinese residents of the neighborhood opposed the downzoning amendment; to reinforce their opposition, they drew on a Lockean variety of liberalism premised on the absolute inviolability of private property.

Even more stunning for the older white residents, however, was this group's use of a communitarian language of the good society that drew on alternative understandings of rationality, normality, community, and what this good society might entail. The recent immigrants, along with some prominent developers, buttressed their opposition by pointing to the history of racially based exclusions in the landscape that were underpinned by zoning and other instruments of supposedly neutral and "rational" land governance through time. They argued that the current downzoning proposal was just one more example of white racism against Chinese immigrants, and a clear circumscription of universal membership in the political *and* economic community of the city.

In the third example, upper-income white residents of west-side neighborhoods contested the removal of mature trees; they used communitarian rhetoric to attack the excessive valuation of private property and *economic* liberalism. They upheld, however, the notion of *civil* or democratic liberalism with respect to their right to protect their interests against individualist profit making, which they represented as counter to the ethos of community. In their promotion of the importance of neighborhood community, however, the neighborhood referent often jumped scales—the trees these groups hoped to save became symbols of belonging to the authentic community of the nation, as expressed through a narrative of roots and associated images of fertility and groundedness. By contrast, the transnational migrants became implicitly associated with rootlessness, and thus ultimately illegitimate members of the local and the Canadian political community.

In a counter-hegemonic effort, the Chinese immigrants drew on the Lockean notions of private property, particularly the sense that nature *becomes* property only as it is reworked for a profit. They reframed the narrative of trees to encompass a positive depiction of transplants and transplantation. Here, transplanted trees and plants evoked a progressive idea of mobility. Rather than the negative association of rootlessness, transplantation represented a sense of new beginnings and of entrepreneurial possibility. In a fundamental if unconscious way, the immigrants called forth a new image of belonging, at a different scale than either the urban or the national; they evoked a community of successful entrepreneurial "transplants" who lived productively *above* space, in the mobile and ethereal world of a spaceless neoliberal global economy.

In all three cases, liberalism was interpreted and reworked by individuals, groups, and institutions who either chose to form new alliances or were unceremoniously thrust into them. In the struggle over spatial hegemony, over the consent to govern the land, older liberal formulations were challenged and new ones produced. While the alliances involved in land governance during the TEAM years drew from narratives of urban livability and national development to bolster both capital accumulation and a socially liberal vision of the city's future, new possibilities and new challenges to this nation format arose with the discourse of globalization and the entry of the transnational migrants from Hong Kong. The ensuing battles over the right to manage urban land thus became wrapped up in questions of territorial identification, political membership, social rights, and the many ways in which these three processes should or should not intertwine.

## FUTURE SHOCK

With the greater visibility of global capital circulation in 1980s Vancouver, accompanied by unpopular urban megaprojects subsidized by the state, the discourse of rational and neutral land

planning began to lose public legitimacy. To bolster that legitimacy, individuals associated with the institutions of land governance—city councilors, provincial leaders, developers, businesspeople, urban planners, urban planning academics, and the mayor—
began a concerted campaign to wrest power back from the new
critical alliances and urban movements that were contesting
neoliberal land policies in the city. Using a number of liberal tools,
including the aura of objectivity and rationality—in contrast to
the supposedly emotional and self-interested discourse of nonpro-
fessionals—they attempted to retain control of Vancouver's urban
environment. One prime example of this effort was the depiction
of all local downzoning or slow-growth movements as selfish and
irrational reactions to the unstoppable (and positive) forces of
neoliberal globalization. This was the case with a zoning amendment put forward by a west-side homeowner in 1990.

John Pitts, a South Shaughnessy resident, spent C$15,000 to
draft a bylaw to rezone nearly two hundred houses in his neighborhood. Bylaw 6694, labeled Preservation District No. 1 by the
residents and sarcastically titled the 'Pitt-Stop' by city planners,
was approved and passed by City Council on July 24, 1990. The
intent of the new zoning was "to preserve the single family usage
and character of the district outlined . . . , with special regard to
encouraging retention of existing dwellings and regulating demolition and replacement of them as a Conditional Use."[5] The new
bylaw reduced floor-space ratio (FSR) to just below the FSR set
by the City Council zoning amendment of 1990; it demarcated
stringent property set-backs specifically designed to end the construction of monster houses in the neighborhood.[6]

One reason planners contested the Preservation District No. 1
name was the implied series (e.g., Preservation District No. 2,
No. 3, etc.). In fact, the use of this zoning freeze as a model for
other districts was mentioned explicitly in the draft proposal of
the amendment under the heading "Potential Precedent for Residential 'Character Area' Zoning." Harry Rankin, of the leftist
political party Committee of Progressive Electors (COPE), said of
the new bylaw: "This is going to be the start of a lot of other
neighborhoods wanting to protect the integrity of their neighborhoods. That is a reasonable demand." But the majority of
councilors did not feel the same way. Gordon Price, an NPA member of City Council, objected to the idea of putting the neigh-

borhood "in formaldehyde."[7] He warned of serious negative consequences for other areas of the city if the bylaw were put into
effect, and blamed the political Left for helping to increase local
area power for cynical political purposes. He told me in an interview in February 1991: "The Left and the Right have both bought
into it [restrictive zoning]. In fact, the Left is more aggressive in
supporting the Pitts because COPE sees their opportunity to build
a constituency on the west side and undermine the traditional
NPA constituency."

He feared the potential loss of a political constituency to the
opposition, but also the threat of the city bureaucracy's losing
"rational" control over land use and land exchange. Despite the
widespread neoliberal rhetoric of government downsizing, a
rationalized, systematic control of urban space by government
agents is crucial to implementing neoliberal urban policy. If local
groups can limit the power of city government to manage land,
and can freeze or remove neighborhood districts from the ravages
of unconstrained development, the circuit of goods and capital
may be hindered, reducing profit and perhaps driving investors
from future city projects. In the case of west-side neighborhoods
in Vancouver, conservative politicians, catering to development
interests that were increasingly international, got caught between
powerful global business interests and forceful local homeowner
associations in their traditional electoral heartlands. As Mike
Davis showed for Los Angeles, the "homegrown revolution" of
the slow-growth movements can divide conservative allegiances
and threaten the political power block of developers and politicians established since time immemorial.[8]

The effects of the slow-growth local control movements in
Vancouver were apparent in the 1990 civic election, where the
issue of neighborhood "community" played a central role. As the
assistant manager of the Economic Development Office put it in
our November 1990 interview: "There's 75,000 RS-1 houses in
the city, and that of course is where the biggest voting block is.
It's not with the community activists and all the business community; it's the homeowners." Recognizing both the extent of
west-side transformation and the degree of angst it was causing
homeowners, COPE, the Civic New Democratic Party (NDP),
and NPA politicians sought to manipulate the territorial connections of neighborhood, community, and political membership

in the society before the November 1990 civic election. In the text of one election flyer targeted at Kerrisdale residents, the Civic NDP wrote: "Under the NPA, big developers and land speculators get free reign, but ordinary people are ignored. Kerrisdale is being transformed—rapidly, and without neighborhood input. It's time for civic leadership that will work with the neighborhood to build for the future."

COPE leaders similarly emphasized neighborhood and the loss of control in the face of unwanted change. They wrote in an eight-page report: "Neighbourhood . . . our neighborhoods make Vancouver great. They make Vancouver work. They are the soul of our city. Under Gordon Campbell and the NPA our neighbourhoods have been shut out, frozen out and sold out. Vancouver is changing and under stress from unequal and unplanned development. We have a city government that listens to the powerful, but not to the neighbourhoods."[9]

Mayor Gordon Campbell and other NPA politicians responded with similar rhetoric, and in discussions, public debates, political flyers, housing forums, newspaper advertisements, and television commercials, the two sides struggled to win political legitimacy as the defenders of territorial integrity at the neighborhood scale. Millions of dollars were spent on the 1990 campaign, and although Campbell retained his position as mayor, COPE, the party traditionally associated with housing and tenancy rights, gained three seats on City Council in the November election.[10]

In mid-November 1990, just before the election, Vancouver's major newspaper, the *Vancouver Sun*, published a seven-part series of articles on current and probable changes in the urban environment. The series, "Future Growth: Future Shock," ran from November 10 to 17 and was accompanied by guest articles, drawings by children, and numerous graphs, statistics, and illustrations. The series was directed by Alan Artibise, the director of the University of British Columbia's School of Community and Regional Planning, and by Michael Seelig, another U.B.C. urban planning professor. Artibise wrote of the series in a promotional article: "What Michael and I have done is an attempt to use our facilities, our knowledge and research abilities to try to contribute to the debate about the future of the region." Seelig wrote, in the same piece: "This is a professor's dream to do interesting applied research that will reach an audience far beyond the academic world. It is immediate, useful and has the possibility to change the face of the region."[11] Why did these professors of urban planning write and edit a series of articles on Vancouver's future with an express interest in changing "the face of the region"? And why was this series published during the week before the Vancouver civic election?

Urban planners, the real-estate industry, politicians, and the media have always participated in the formation and promotion of urban development and the ideas associated with particular neighborhoods and landscapes. As with the earliest forms of control over the city's neighborhoods, this consortium of urban actors has sought to retain the authority and legitimacy to decide how and in what ways the city should develop. They attempt to retain this authority through the discourse of "rational" planning, as well as through the ongoing manipulation of distinction and social identity. Maintaining authority in land-use planning is central to the ongoing goal of capital accumulation; it also encompasses issues related to the role of the state and the legitimacy of a particular narrative of liberalism and national identity in Canada during a period of rapid restructuring. In this broad scenario, the "Future Growth: Future Shock" newspaper series is one thread in a tapestry of neoliberal hegemonic formation concerning the urban future of Vancouver.

Soon after the Pitts amendment passed, politicians, planners, and developers met to determine the next plan of action. According to Artibise, the "Future Growth: Future Shock" series was outlined following a meeting of Vancouver developers who had gathered to articulate new strategies in the wake of the increasingly active participation by local citizens in urban policy making. He said in an interview in Vancouver in November 1990 that developers were particularly upset by what they termed increasing "NIMBYism" (Not in My Back Yard), a mentality the developers depicted as selfish individualism. After the meeting, some of the developers approached Artibise to write a series of articles similar to one they had seen by the architects Neal Peirce and Curtis Johnson in the *Seattle Times* on the future development of Seattle's Puget Sound region.[12] The *Seattle Times* series stressed the inevitability of population growth and the need for coordinated and planned development in the face of rapid regional expansion. As "no one would believe [the articles] if a developer wrote them," Artibise told me, the planning professors were commissioned on a contract basis to write a similar series for the *Vancouver Sun.* According to Artibise, the series was a "calculated attempt to change an ideology."

One of the key narratives that Artibise and Seelig drew from in their series was a communitarian philosophy that criticized excessive individualism and its potential for cultural and political anarchy. The ways in which the authors borrowed from this philosophy enabled the conjoining of Canadian narratives of community (versus excessive individualism) and authority (versus anarchy) with the need for rational urban control. The series discredited supposedly chaotic and individualist urban movements by private individuals and groups, and depicted more positively the holistic and professional vision of planners and politicians, and the neutral, rational, efficient workings of the (global) market.

The authors used various forms of ideological persuasion to win consent for their neoliberal, pro-growth agenda. These included an effort to *universalize,* as common to all, the values and interests of those represented in the articles; to *rationalize* these interests as logically consistent; and to *naturalize* the beliefs as self-evident—part of the general commonsense of the society.[13] For example, efforts to rationalize the interests expressed in the articles as logically consistent were evident in the strong emphasis on the inevitability of regional growth. The series heralded the changes brought about by increased global integration as positive and inescapable. Stanley Kwok, the vice president of Concord Pacific, wrote in one article: "The Vancouver of 2010 will largely be shaped by global forces, which are beyond our control."[14] The provincial premier, Bill Vander Zalm, said in an interview that was quoted in the series: "It is impossible to stop growth. We are part of a global community."[15] Population growth, increased housing demand, densification, growing Pacific Rim linkages, and economic expansion were shown to be unavoidable processes, and attempting to curtail them foolish and childish. Artibise and Seelig wrote in the first article of the series: "Lock the gates. If the choice were ours, we would stop people from moving into the Pacific Fraser region. . . . We are the Peter Pan of urban regions. Like Peter Pan, we would simply like to declare: 'We won't grow up.' But locking the gates is not a realistic option."[16] Opposition to growth was thus positioned as childish in the extreme, as irrational as Peter Pan, who refused to grow up and accept the responsibility of (rational) adulthood.

The persuasive rhetoric of reason continued with a number of articles that stressed the positive influence of unfettered market forces. These forces were projected as inherently rational, and economists who supported more freedom for the market were positioned positively against others who argued for zoning controls and other restrictions. For the article on November 12, Artibise and Seelig interviewed David Bond, a vice president of the Hong Kong Bank of Canada. This economic highflyer decried Vancouver's zoning policies for "delaying the market." Later in the article, the authors reiterated the inevitability and benefits of growth and densification alongside neoclassical assumptions associated with the positive effects of freer markets. These assumptions were written as facts, separated from the other paragraphs in the article by marked indentations.[17]

In addition to the reputedly positive effects of densification for urban residents, Artibise and Seelig affirmed in a later article that developers would profit from the changes as well. According to the authors, the profits accorded developers as a result of allowing greater densities would result in better and cheaper facilities for residents and buyers. For example, they wrote,

> These changes need not mean increased costs to the public. By allowing developers to build at much higher densities, the public is in fact making a gift to these private developers and land owners. The gift is a substantial increase in the value of their land. The bulk of this extra "profit" from increasing the allowable building density will be used to ensure the necessary facilities are ready when the owners and renters move in. The payoff is a better residential community for buyers, renters and developers.[18]

The use of a number of different voices in the Vancouver community heralded the effort to universalize the perspectives offered in the series. Numerous nonjournalists—mostly urban planners, developers, and bankers—were invited to write short guest pieces in which they described their personal visions of Vancouver's future. In a few articles the attempt to paint urban growth and densification as a common desire was made through linkages with other "dense" regions. The "universal" goals of Denmark, for example, were identified with those of Vancouver, with the homogenizing statement that, "in Denmark, single-family homes are white elephants. People prefer townhouse and apartment-style developments to single-family homes."[19]

Thinly veiled threats accompanied the articles. If city residents persisted in their irrational and selfish attempts to control the local environment and to slow market forces, then business would move elsewhere. Furthermore, since increased density was so vigorously contested in Vancouver, the precious agriculture of the Fraser Valley would have to be sacrificed to ongoing urban sprawl: "One result of this NIMBY-NIMTO [Not in My Term of Office] pressure is that developers are moving further up the valley, where municipalities are still hungry for development. A billboard along the Trans-Canada Highway, for example, proclaims: 'Chilliwack—Open For Business, Call the Mayor.'"[20]

The flexibility of capital to move quickly and seek out the most profitable investment sites was a frequent theme throughout the series. It was made clear that Vancouver residents who refused, like Peter Pan, to grow up and accept the numerous changes in the region could be condemning their city to a future without a viable economy.[21] The frictions characteristic of slow-growth neighborhood movements would undermine the city's attractiveness as a site of investment for global investors and permanently injure the regional economy. The authors warned in the final article of the series: "As the region grows in international importance, the traditional boundaries between countries become increasingly meaningless. How we deal with that reality will largely determine our future."[22]

The success or failure of efforts to transform older discourses such as that of social liberalism depend on how well they can tie into preexisting, commonsense understandings in society.[23] Here I focus on the commonsense understanding of the particular narrative of Canadian liberalism that relies on respect for authority, modern faith in liberal reason, and its distinction from constructions of "excessive" individualism associated with the United States. The widely shared philosophy of this kind of liberal-communitarian composite ties together some of these strands and remains an important touchstone and source of national identity for many Canadians.[24] So how were efforts to form a popular consensus in support of *neo*liberal reforms articulated with this particular national narrative?

In a Canadian context, communitarianism is often positioned as a critique of the excesses of individualism, including those individual rights and freedoms which appear to undermine social coherence.[25] Community, defined empirically and geographically, is foregrounded, as are notions of the community's "common" good. Many of the articles by Seelig and Artibise joined in the critique of a classic, rights-based liberalism associated with negative American values. They targeted civil individualism and cultural and political anarchy, as well as the "chaos" of popular democracy and the corrosive effects of excessive antagonism. The warnings against the dangers of excessive individualism and lack of authoritarian control were most clearly expressed in a November 13 article entitled simply "Politics." This article—which began: "Who's in charge? No one and everyone"—featured a large graphic under the caption: "Vocal citizens band together to force politicians to act from a local rather than a regional perspective." In this graphic, an angry gesticulating mob carried a sign that read "VOTE," and pointed toward a series of arrows that led to an explosion. Each figure appeared to be acting alone within the throng. The unrestrained democratic energy of the crowd was depicted as regressive and destructive, the harmful result of a lack of central authority. Mouffe writes of a neoconservative movement's similar dramatization of democratic chaos in the United States at the end of the 1960s: "Raising the spectre of the 'precipice of equality,' this group, composed of prestigious intellectuals united around the reviews *Commentary* and *The Public Interest*, launched an offensive against the democratic wave which the various social movements of this decade represented. They denounced the excess of demands that this multiplication of new rights imposes on the state and the danger that this explosion of egalitarian claims poses to the system of authority."[26]

The Pitts zoning amendment in South Shaughnessy was singled out in the series as a particularly selfish attempt to wrest planning control away from more objective and rational public agents:

Consider the following. Fed up with what they saw as the destruction of their Shaughnessy neighborhood through demolition of houses and construction of "monster houses," John Pitts and his neighbors hired their own architects and lawyers to prepare a bylaw to preserve the area. Nicknamed "the Pitt-Stop," it marked the first time citizens had drafted their own bylaw, sent it to city hall and convinced council to adopt it. What would happen if every neighborhood had its own bylaws?[27]

To squelch this growing civil rights–based liberalism, the authors implied that the chaos of popular democracy must be countered with the authority of government and an ideal conception of the good society. Drawing on this older tradition of civic republicanism, Seelig and Artibise wrote: "A growing number of planners, developers and politicians are saying that popular sovereignty has become a euphemism for abandoning responsible, representative government. . . . Citizen crusades can be dangerous because they do not focus on the broader issues and tend to dismiss the rights of the larger community."[28]

The broad appeal to communitarian traditions thus critiqued liberalism on the grounds of spatially based notions of community and the common good, alongside the discreditation of dissidence.[29] So-called excessive individualism, such as manifested in the NIMBY and NIMTO movements, was positioned as fracturing larger-scale community interests and ultimately leading to harmful regional effects for the environment and for business.[30] The main tenor of the attack relied on a negative depiction of political and cultural anarchy embodied in excessive individual rights. Further, it was the excessive *civil* rights of democratic liberalism, those of free speech and association, rather than the *economic* rights of private property, that were projected as the most problematic vis-à-vis the good of the wider community.

The promotion of conservative civic republicanism in the articles was also evident in efforts to harness nostalgia. Vancouverites were situated as members of a historically rooted community that could be revived through a return to the "premodern," even as the city embraced the authors' visions of global integration and rapid urban growth. This nostalgia was engaged by linking a fin-de-siècle lifestyle of community closeness with increased urban densities.

"You want a picture of the region in 20 years?" says Michael Goldberg, head of Vancouver's International Financial Centre. "Look back to 1900 and you'll see it all. . . . At the turn of the century people lived in apartments above shops and travelled by streetcars that stopped at every major intersection. The escalating cost of land, rising gasoline prices, environmental concerns and gridlock will likely force us into this new-old lifestyle. But instead of being grim, this future of living closer to each other could bring with it some of the lost friendliness and closeness of urban life. Imagine getting up in the morning and walking to the corner store to get fresh bread for breakfast. On the way, you greet a neighbor who is walking to his office. Children are walking and bicycling to school."[31]

Children were a common theme throughout the series, which indicates how the past and future were carefully sutured together in these neoliberal urban visions of warm planetary integration. Children were portrayed as representing the future of the region, but they also embodied a sense of historical rootedness, of tradition, stability, and domesticity. Children's drawings featured prominently alongside a number of the articles. Many depicted futuristic images such as skyscrapers and new transportation systems juxtaposed with the city's familiar mountains. All seemed to espouse a willingness, indeed an eagerness, for a new kind of urban environment. Unlike the citizens who refused, like Peter Pan, to grow up, the children were portrayed as rational and unsentimental about their future and the future of the region.

Following the publication of the series, a number of letters to the editor, published in the *Vancouver Sun* on November 26, 1990, indicated some of the vast resistance to the liberal-communitarian narratives employed by the authors. Many writers showed a deep understanding of the authoritarian implications of opposing democratic individualism. Michael Sims wrote: "The authors say we must relinquish control over our neighborhoods to planners—smart people such as themselves. Sounds a bit self-serving to me. Sorry that they find democracy troublesome, but I think we should try to hang on to it."

Others indicated awareness of the clear connections between the discourse of globalization and growth that was promoted in the series, and the potential profitability of these processes for a select few. R. W. Fearn wrote: "I wish Arthur Erickson and *The Vancouver Sun* would get off the 10-million-people-for-Vancouver bandwagon. You both have a vested interest in that kind of growth, but your attitude has profound implications." And Andrew Roberts wrote in the same vein: "According to the authors of your series, densification is the greatest thing since sliced bread. They concede that most residents oppose increased density, but quote the vice-president of the Hong Kong Bank of Canada as saying 'we need to force changes in density.' Oh really? Who is going to force whom to do what? Does the *Sun* endorse this view? I trust *Sun* editor-in-chief Nicholas Hills will see that the views of 'Peter Pans' also get some coverage."

Social resistance to the neoliberal changes in Vancouver blocked the free flow of global capital into and through the city. The various institutions of land governance, aided by businesses invested in the ongoing growth of the city, including the *Vancouver Sun*, sought to counter this resistance and promote the city as an effective site of international capital accumulation. However, for the policies and ideological promotions of these institutions to remain viable, they had to articulate with preexisting assumptions in Canadian society. Developers, bankers, realtors, conservative academics, and municipal politicians thus fought to secure urban neoliberal policy by manipulating cultural narratives of the socially liberal nation, particularly those associated with the nostalgic image of a well-planned, obedient, coherent, and unselfish Canadian community. As always, this effort to produce hegemony was uneven in its effects, and the struggle over the right to control land continued through several more highly publicized conflicts.

## ZONING AND LIBERAL DEMOCRACY

One of the major landscape controversies in Vancouver took place during a series of public hearings on downzoning in the neighborhood of Shaughnessy Heights in 1992. The meetings were intended as a public forum to discuss land governance, and to demonstrate the importance accorded public sentiment in the area of land management. The discussions, however, quickly became linked to broader questions of rights, representation, access, community good, and the interpretation of the public sphere of contemporary Canadian liberalism.

Before I turn to this set of hearings, however, I want to spend some time examining the history of zoning in Shaughnessy Heights, and why the stakes over this particular zoning struggle became so high. Despite the persistent discourse of neutrality and rationality in the workings of the market (vis-à-vis the buying and selling of land) and in the practices of land governance, historical zoning patterns in Vancouver had been bound to racial and class-based definitions and processes since the arrival of white settlers in the city. The central players in shaping urban form were (and have largely remained) the executives and politicians connected with the Canadian Pacific Railroad (CPR).[32] Through the selective and targeted buying and selling of land, alongside the production of specific images that pertained to that land, railroad executives had maintained exclusivity and promoted a distinctive, class-based, racially coded lifestyle in every neighborhood of Vancouver since its establishment in the nineteenth century. This privilege and exclusivity was particularly evident in the formation of Shaughnessy Heights.

The original terminus of the CPR was extended from Port Moody to Coal Harbour (located in present-day Vancouver) after deliberations between railroad executives and politicians. To encourage this twelve-mile extension, government officials promised the CPR two Crown land grants near the new terminus. On February 13, 1886, 480 acres in the Coal Harbour reserve and nine square miles of forest land south of False Creek (5,795 acres) was given to CPR executives Donald Smith and Richard Angus. This "untouched forest land" now contains the neighborhoods of Shaughnessy Heights and Kerrisdale.[33]

After the decision to locate the terminus in Coal Harbour, a link with Asia was promoted by CPR executives, who advertised Vancouver as an exciting new international center. Vancouver's expected emergence as the "gateway to the Orient" drove up land prices in the city and fueled the early speculative rush that was so lucrative for CPR coffers. At the same time, the company portrayed itself as one whose imperial significance and duties outweighed the mundane business of transportation. According to CPR propaganda, the railroad-steamship connection was the new, "all-British, all-red" route to Australia and the riches of the Far

East, and should be considered and represented as upholding British national and imperial interests and ideals around the world.[34]

The gross earnings of the CPR during the first period of great land speculation grew from C$13,195,000 in 1888 to C$29,230,000 in 1899; by the end of World War I, the company's earnings had mushroomed to C$157,357,000, most of it garnered through real-estate speculation and property development.[35] The considerable economic power wielded by the railroad during these years allowed the CPR to play a major role in determining Vancouver's street layout and general land-use patterns. For example, L. A. Hamilton, the CPR surveyor and city land commissioner in 1885 and 1886, planned all of Vancouver's major east-west avenues and cross streets. This kind of dual role between business and government was common. From 1886 to 1897, three mayors were actively involved in real estate, and many City Council members were either CPR executives or wealthy businessmen.[36] Indeed, the major stipulation for election as city councilor or mayor was property ownership; an amendment to the Act of Incorporation of 1887 required the mayor to possess property worth C$2,000 and city councilors C$500 to run for office. In some wards this limited the field to three or four men. Unsurprisingly, proposals to tax land were consistently voted down.[37]

In addition to dictating urban form, CPR surveyors and city mayors also arranged how the city was to function socially and aesthetically. Clearly delineated boundaries corresponded with the location of CPR property in the West End and to the south of False Creek. Residential lines that marked the divisions between the wealthy, the white, and the socially acceptable matched the holdings of CPR executives and businessmen in the western end of the city. Since CPR executives formed the core of the social elite in Vancouver, where they chose to live soon became the most prestigious and exclusive residential area.

Protecting the racial and class exclusivity of the West End was doubly advantageous for wealthy railroad executives. Their personal fixed capital interests in house and estate were secured, and their extensive land holdings in the area were guaranteed to remain attractive for future subdivision and sale as residential areas.[38] The exclusivity of the West End was maintained through the discretionary use of various instruments of planning. Fire and health regulations were invoked to complement private covenants, and the negotiation of building loans, sale of finished homes, and securing of mortgages were scanned to ensure a buyer's proper "qualifications" for living in the area. In this manner, Vancouver's property industry was able to exert considerable pressure on land development long before more formal city zoning measures were passed.[39]

The informal zoning procedures that shaped urban residential development also shaped the taste of potential home buyers. Real estate developers accented exclusivity in advertisements and manufactured desire for selected neighborhoods based on image. Part of the image of these suburban elite neighborhoods revolved around the exclusion of racial, ethnic, or class groups other than wealthy British homeowners. This image was maintained and protected by the selective enforcement of informal zoning measures such as fire zones and health codes that were implemented to exclude Asians and other "undesirables" from the area.[40] Representations of cultural difference were frequent and were naturalized as essential racial characteristics—thus warranting the exclusion of members of those "races" who did not fit the image of the neighborhood.[41]

In 1908, fearful that they might be sucked into the vortex of the immigrant working-class neighborhood of South Vancouver, the people living west of Cambie Street and south of False Creek seceded from their neighboring area and formed the municipality of Point Grey. (This area now encompasses the local areas of West Point Grey, Dunbar Southlands, Arbutus Ridge, Kerrisdale, and Shaughnessy Heights; see Map 1 in Chapter 2.) The area was still mainly undeveloped CPR land from the original Crown grant of 1886. When the CPR subdivided this land for development and sale, company executives joined with politicians in designing distinct neighborhoods for different elements of the population. Lots in Fairview, located just south of the industrial area of False Creek, for example, were auctioned off publicly in 1890 and were purchased in bulk for future speculation and subdivision. South Vancouver, a working-class suburb located just east of the CPR land, was also the site of tremendous speculation. Point Grey, in contrast, aided not just by incorporation as a separate municipality but also by a deliberate marketing strategy of exclusiveness, was developed and heralded as a carefully planned and highly regulated residential district from its inception.

Point Grey's incorporation as a separate municipality was promoted by the Point Grey Improvement Association (PGIA), an organization composed of property owners frustrated by their inability to develop their land for maximum profit. After the PGIA lobbied successfully for secession from Vancouver in 1908, the newly formed council built paved roads, set aside money for parks, and adjusted municipal taxation to discourage speculative land purchases in the area.[42] Similar actions served Shaughnessy Heights, a small section of Point Grey still completely owned by the CPR. In 1909, before selling lots, the CPR installed water, sewer, and other utilities in the property and constructed winding, elegant high-quality roads. When the land was finally released, it was not sold at once but in strategically timed amounts to maintain market value. Lots were sold in large parcels with a number of stipulations about required size, building quality, and house cost to ensure a minimum of speculative activity.

In addition to these early subdivision and planning measures, Shaughnessy Heights was further set apart by the Shaughnessy Settlement Act, passed by the provincial legislature in 1914. This act "recognized and sanctioned the undertakings of the CPR to ensure Shaughnessy's status in the City," thus legislating the informal sanctions already in place.[43] Further restrictions were placed on neighborhood development via the Shaughnessy Heights Building Restriction Act of 1922. This act removed zoning control from the municipality and prohibited further subdivision of the land. Conditions for land use in the area included the stipulation that no buildings could be erected on the land "which should be a nuisance to the person or persons owning other land in the area."[44] The CPR placed further restrictive covenants on Shaughnessy deeds by ensuring that only single-family homes could be built.[45]

Point Grey land was developed with the expectation of long-term profit predicated on the ability to establish and maintain the neighborhood's desirability. This desirability was established by the provision of high quality infrastructure, much of which was paid for by provincial and municipal governments; it was

maintained through the selective enforcement of private covenants and zoning regulations that upheld racial and class exclusivity. High profitability in the area induced further infrastructural improvements and municipal money in a snowball effect.
As the promise of profit in Point Grey grew, so did the interest
of investors and developers in even fancier projects, parks, and
plans. B.C. Electric Railway Company (BCERC), for example,
opted to provide tramway service to the Point Grey area long
before the population level warranted it. As of 1911, the land contained only 854 dwellings; a year later, it boasted sixteen miles
of tramways.[46] The famous planner Harland Bartholomew wrote
of Point Grey in 1928: "The attractiveness of Point Grey as a residential district is due in part to the willingness of municipal
authorities there to set aside public park and play areas. Point
Grey has character. It attracts a class of citizens who want parks
and are willing to pay for them. The parks bring new citizens,
and they in turn create additional property wealth which can be
taxed to pay for lands already acquired and new areas needed to
serve the new population."[47]

The interest in Point Grey as an exclusive suburban development continued through the interwar years, and the controlled and
carefully defined borders between communities became further
established. By the mid 1920s, the West End and the east side had
become the province of the middle and working classes, while
West Point Grey, Kerrisdale, and Shaughnessy housed the highest-
income groups. Up until World War II, east-side and downtown
neighborhoods were ethnically mixed, while west-side neighborhoods remained nearly 100 percent white and Protestant.[48]

Municipal acts, town planning, private covenants, and zoning
measures in Vancouver were dictated and manipulated by local
property owners and developers to maintain this pattern of neighborhood exclusivity.[49] The Vancouver Property Owners' Association (later the Associated Property Owners), representing the
CPR, the B.C. Land and Investment Agency, and the Vancouver
Land and Improvement Company, was a particularly powerful,
extremely conservative force in Vancouver civic affairs. This association attempted to mask its vested interests in selective urban
growth by representing itself as a "non-political, fact-finding
agency."[50] The organization lobbied heavily for a provincial Town

Planning Act, which was passed in 1925, and then teamed up
with the Vancouver Real Estate Exchange and the Board of Trade
to lobby the provincial government for municipal zoning. When
planning commissions were set up and zoning boards organized,
many of the participants or leaders of these new city organizations
also held positions as large property owners and executives of
real-estate firms.[51]

Point Grey was amalgamated into the City of Vancouver in
1929, but until then functioned as a separate municipality, with
its own planning covenants and zoning bylaws. In 1922, the council passed the most comprehensive zoning regulations in all of
Canada for the new area of Shaughnessy Heights. Long before
any major development was undertaken, the potential uniqueness
and attractiveness of Shaughnessy were made clear and maintained through the actions of the Point Grey City Council, which
"insistently held to the idea that they were put into office to
carry into effect the ideals held by the residents that Point Grey
was to be developed essentially as a first-class residential district."[52] For the next six decades, this "first-class" district managed to protect its highly privileged and exclusive character
through a series of zoning measures designed to block unwanted
change and unwanted arrivals from outside the spatially defined
"community."

Although Shaughnessy Heights was already protected by strict
zoning covenants, these regulations became even stricter in the
1980s in response to an increasing threat of change. In May 1982,
after a background report was prepared by the Vancouver Planning Department (at the instigation of the Shaughnessy Heights
Property Owners Association [SHPOA] and at municipal expense),
City Council rezoned the area from RS-4 (One Family Dwelling
District) to FSD (First Shaughnessy District).[53] On the same day,
council approved a series of "design guidelines" for the area,
amended the subdivision bylaw, and established the First Shaughnessy Advisory Design Panel. Controls over land use were outlined for the area in a special zoning regulation entitled First
Shaughnessy Official Development Plan By-law No. 5546.[54] The
design panel served as an additional level of protection against
unwanted development in the area. Members of this design panel
were appointed by the City Council and included two represen-

tatives of the SHPOA, two residents of the neighborhood, one representative from the Architectural Institute of British Columbia, one from the B.C. Society of Landscape Architects, and one from the Heritage Advisory Committee.

Panel members reviewed all proposed developments in the area with respect to both the original Shaughnessy bylaw and the First Shaughnessy Design Guidelines. These guidelines were extremely detailed, including minute architectural design features such as the preservation of roof, entrance, and fire escape treatments, as well as broader principles of massing and siting.[55] In the General Application section of the 1982 manual (an addendum to the earlier development plan), "neighborliness" was emphasized in the first paragraph, and "sense of timelessness" in the next. Other sections called for "authentic appearance" and "quality." In each, the elements of an extremely wealthy, firmly established, and distinctly unchanging character were fiercely and categorically outlined and defended. For example, in the allowable "materials" subsection was the following paragraph:

> One of the prime criteria for establishing the acceptability of exterior materials is based on quality. High quality materials are defined by their cost as well as other aspects noted below which substantiate their use. *Sense of Timelessness*: The materials retain their intended shape for many years from the time of installation and keep this shape without wrinkling, buckling, or curling. . . . *Substantial Structural Qualities:* The material has an intrinsic structural quality, and gives the appearance of a certain structural resistance. It does not have the appearance of a superficial, "pasted on" element. . . . *Authentic Appearance:* A principle of building which has for the most part been recognized in Shaughnessy, involves the "honest" use of materials, avoiding the use of impermanent, low-quality materials which masquerade as more enduring, architectural ones. The materials in question should avoid imitating others.[56]

In each paragraph, a supreme paranoia regarding displacement—particularly by fraudulent, imitative, masquerading, and non-authentic, non-Shaughnessy types of buildings—reflected the fear of losing the high distinction and stability established for the neighborhood.

As a result of the increasing pressures of development, exacerbated by the city's integration into global networks of capital and migration, however, the multiple layers of protection established for the neighborhood through time became increasingly ineffective. Resistance to the ensuing changes took many forms, including long, angry letters to City Council and to the editors of local newspapers. Conflict associated with the transformation of the neighborhood also began to occur in public spaces such as meeting halls, City Council chambers, neighborhood property rights associations, urban planning sessions, and media outlets. In the public hearings of 1992, older residents and their allies, and newer residents from Hong Kong and their allies, clashed over contemporary practices of land governance, drawing on differing understandings of liberalism and community to bolster their respective arguments about urban, spatial rights.

## THE HEARINGS OF 1992: REWORKING THE PUBLIC SPHERE OF LIBERALISM

In a series of six public hearings on the topic of downzoning in South Shaughnessy Heights, the polarization of the long-term white residents and the newer Chinese residents became immediately apparent. A majority of primarily white SHPOA members advocated restrictions on future redevelopment in the area. Following this, a group composed almost entirely of recent Chinese immigrants from Hong Kong quickly formed an ad hoc committee to oppose these proposed development restrictions. This committee, which later convened as the South Shaughnessy Property Owners' Rights Committee (SSPORC), campaigned strenuously against the downzoning. Aided by a number of developers, most prominently Barry Hersh, a homebuilder and president of the West Side Builder's Association, the campaign alleged racism to discredit the SHPOA and its followers.

After the first hearing in September, the SSPORC sent leaflets to Chinese homeowners in the area, urging them to ask Chinese friends from the lower mainland to attend the rest of the hearings in a show of public alliance. These leaflets were written in Chinese, whereas the original survey questionnaire on the topic of downzoning had been written in English. At the public hearings, Chinese speakers complained that the survey, distributed by the Vancouver Department of Planning, had not been translated into Chinese, reducing the amount of information gathered to make informed decisions about the proposed change.[57] Some of these speakers spoke Cantonese at the hearing and used an interpreter. At several points, they were heckled by white members of the audience for not speaking English.

The widespread publicity of this racial polarization (the hearings were televised and reported nationally), indicated new stakes in the battle over urban design and control of the Vancouver landscape. Public hearings such as these are generally presented as models of open democratic debate and spaces of rational decision making; they are the cornerstones of the public sphere of liberal theory, helping to support the ideology of an ever-widening sphere of inclusion and universal membership in the political community. But despite the discursive rationale of openness and inclusion in these types of hearings, they rarely function this way. In most land-use hearings, for example, as noted in the discussion of Shaughnessy's landscape history, the outcomes concerning appropriate land use have historically led to greater *restrictions* on neighborhood membership. Through time, these types of territorial restrictions affect both the material and the discursive constructions of neighborhood belonging and cultural authenticity, and reflect back in a mutually constitutive cycle on the debates and outcomes of political processes such as the land hearings.

The public hearings in Shaughnessy Heights thus represented the first time that a radically different interpretation of community belonging and of the good society was promoted by a group not composed primarily of wealthy white residents. For the first time in the city's history, a fairly large number of nonwhite entrepreneurs moved to neighborhoods formerly protected from unfamiliar outsiders by virtue of their high prices and exclusive zoning covenants. This outside group was able to form alliances with local developers for a pro-growth agenda that threatened the carefully established symbolic value of west-side neighborhoods. By raising the spectre of racism, which could be easily buttressed by the historical evidence of racism in earlier zoning measures, the Hong Kong Chinese and west-side builders defeated the SHPOA effort to restrict further development in the neighborhood. As a further warning to their white opponents, some members of the SSPORC publicly threatened to write Chinese friends and contacts in Hong Kong and worldwide, reporting that Vancouver was a racist society and advising against further (global) investment in the city.

In defense of their position against downzoning, many Chinese speakers extolled the virtues of a superior Chinese way of life. Rather than contesting the rhetoric of cultural difference, Chinese homeowners in Shaughnessy used it to their advantage, lecturing enthusiastically about extended families, respect for the elderly, communal closeness, education, and hard work—in contrast to their representations of lazy, unfamilial, and undemocratic white Canadians. Similarly, they appropriated the meanings of racism and nationalism to their advantage, decrying English-only language survey forms as inherently intolerant and hence un-Canadian, and neighborhood rezoning proposals as manifestly racist.

The understanding of what constituted the good society became a major source of contestation. Definitions of what it meant to be Canadian or to live in the neighborhood of Shaughnessy Heights were interrogated as racially based cultural constructions rather than natural affinities. But rather than simply contesting white definitions of Chineseness and of appropriate Chinese places to live, the financial power and cultural savvy of the immigrants gave them the power to completely invert this process of racial and spatial definition and present dominant Chinese definitions of whiteness.[58]

The new Chinese homeowners in Shaughnessy allied themselves with developers, realtors, and politicians eager to join in new increasingly global business ventures. This slight shift of the economic and political power in the neighborhood allowed the public voicing and material representation of white values and norms from the outsider's perspective. The spaces of rational discourse, an integral component of the public sphere of liberal theory, were thus employed to contest white Canadian cultural assumptions of appropriate community values and membership, which had operated spatially to maintain white hegemony in the neighborhood since its inception. Through the invocation of history and geographic context in this process of contestation and inversion, some of the actual spaces of the liberal public sphere (public hearings on land use), were also exposed as traditional spaces of hegemonic production for a dominant group.

To give final weight to this inversion, many Chinese residents voiced their concerns in Cantonese, deliberately employing the language of popular democracy in defense of their individual and economic rights. They further spoke of these rights as integral to the Canadian liberal state. In the hearings, much to the dismay of their white Canadian listeners, Chinese Canadian speakers invoked notions of freedom, family, individual rights, property rights, and democracy itself in defense of their position. One Chinese Canadian speaker said (through an interpreter): "I live in Shaughnessy and we built a house very much to my liking. The new zoning would not allow enough space for me. . . I strongly oppose this new proposal. Why do I have to be inconvenienced by so many regulations? This infringes my freedom. Canada is a democratic country and democracy should be returned to the people."[59] The threat to established norms that this position heralded became clear when I interviewed a SHPOA member in Vancouver in November 1992. She said to me: "I can't describe to you how it feels to be lectured to about Canadian democracy by people who have to use an interpreter. . . . Lectured about laziness, how we don't need big houses because we don't take care of our parents like the Chinese do. . . . I cannot understand how *anybody* could go to another country and insist they can build against the wishes of an old established neighborhood. The effrontery and the insolence takes my breath away."

Historic zoning patterns and landscape struggles in Shaughnessy Heights give some indication of how institutional tools such as urban planning were implicated in the social reproduction of a dominant white elite in Vancouver. The contemporary conflict over downzoning also demonstrates how, by using alternative assumptions about the nature of community, property, and family, the recent Chinese residents contested this form of institutional hegemony. In their ostensible challenge of a local zoning policy, the new residents exposed some of the fault lines of liberalism, especially the culturally inscribed assumptions of spatial belonging and its connections with political membership.

One of the key underpinnings of the liberal public sphere is the assumption of neutrality and "spacelessness," that is, the absence of space as a marker of personal identification.[60] In the ideal conceptualization of the liberal public sphere, the citizen is able to speak "as if" he or she is on an equal level with all other citizens, regardless of personal status. The public sphere is represented in modern liberal theory as a neutral arena where rational deliberations are made that are distinct and separate from the nonneutral spaces of the private realm.[61] In this conceptualization, social inequities are not an impediment to participation, since the public sphere functions as a neutral, status-blind, space-free arena of deliberation.

In actual public spaces, however, numerous scholars have shown how informal obstacles to participation can remain long after formal exclusions are removed.[62] These obstacles are frequently cultural and economic rather than legal. For example, a distinctive culture of civil society is based on norms of rationality or order or familiarity that are implicit in the process of bourgeois class formation. Thus, the cultural markers of status infect deliberation, regardless of legal or formal "neutrality." "In stratified societies," as Fraser writes, "unequally empowered social groups tend to develop unequally valued cultural styles. The result is the development of powerful informal pressures that marginalize the contributions of members of subordinated groups, both in everyday life contexts and in official public spheres. . . . Subordinated social groups usually lack equal access to the material means of equal participation. Thus, political economy enforces structurally what culture accomplishes informally."[63]

Given the historically unequal participation of subordinated groups such as women, racialized minorities, and the working class, the ideas of the dominant group are generally normalized and reproduced as neutral in the context of liberalism. They are then used to justify state policy. With the Hong Kong immigrants' arrival in Vancouver, however, this normalizing process was challenged. The Chinese immigrants who could buy property in Shaughnessy Heights were a dominant class fraction with a distinctive set of cultural markers. As members of an established class, this group had the finances, the self-confidence, and the cultural capital to promote new forms of cultural and economic distinction in the landscape. They were able to enter public meetings and voice their differences and their desires. In this way, they successfully contested the assumed connections between high culture and a landscape in the tradition of a white British elite. They also disputed the privileging of the landscape in terms of "character" and of use value over "profit" or exchange value. Through the allegations of historic institutionalized racism in zoning and urban design in Shaughnessy Heights, the Hong Kong

Chinese ad hoc committee opened up the implicit connections between landscape "character" and the ongoing hegemonic reproduction of a white Canadian elite.

Another assumption of the liberal public sphere the new Shaughnessy residents contested was the necessary separation between the public and the private. For most liberal theorists, this separation is essential for the public sphere to function; without it, private interests can dilute the rational discourse necessary for a common consensus on public issues. Numerous feminist critics have shown, however, how notions of the private and the public are cultural classifications inherently bound up in relations of power. The labels of private and public are frequently deployed "to delegitimate some interests and valorize others."[64] In the past, invoking privacy often functioned ideologically to delimit the boundaries of public discourse, particularly in the realm of private property.

With the public hearings on downzoning in Shaughnessy in 1992, the "private" subject of racism was brought into public focus. The issue, like that of domestic violence in preceding eras, was considered a private affair best addressed on a case-by-case basis. The normal boundaries of public and private were thus transgressed as the Hong Kong Chinese residents in Shaughnessy claimed that the downzoning amendment was functioning as a screen for racist desires to exclude nonwhite people from the neighborhood. In this case, the unspoken norm of a predominantly white neighborhood was exposed and challenged through the refusal of existing notions of appropriate public discourse. In an ironic twist, the owners of *private property* invoked their dominant class position to *open* public discourse rather than the reverse.

The new Chinese homeowners also publicly pursued notions of the good society that diverged from established neighborhood norms. By emphasizing what they represented as Chinese values of filial piety, respect for education, and hard work, and by contrasting these with their perception of white values, the recent immigrants challenged the formerly implicit neighborhood understandings of the good life and of the best future for the society. In the ideal public sphere of liberal theory, notions of the good life belong in the private realm; historically, however, these public-private distinctions have been shown to normalize dominant values. Questioning the assumptions of an appropriate Shaughnessy landscape founded on the image of pastoral England also exposed the construction of a racially defined and spatially secured dominant elite.

In this manner, the assumptions of a public-private split and a neutral "spaceless" territory were shown to mask a cultural norm that benefited a dominant white elite in Shaughnessy Heights. This norm was reproduced through a process of consent essential in the ongoing production of hegemony. The arrival of a culturally sophisticated group which did not "consent" to the cultural norms of the Shaughnessy landscape disturbed the moral and cultural leadership the dominant group needed to maintain this hegemony. With alternative interpretations of the common good aired publicly in the battle over downzoning, the heretofore normalized conceptions of the landscape were shown to operate historically in the interests of a dominant, racially exclusive class. More importantly, the struggles that emerged with new forms of transnational articulations were not merely over particular public spaces, but over the political practices of liberalism itself. In the following section, I address a third and final case in which liberalism, landscape, and political membership were intertwined.

## ROOTS AND TRANSPLANTS: RECODING CANADIAN COMMUNITY

There is probably no better symbol for British Columbia at this time than trees. Nice, big, old trees. It was through the mass destruction of hollies and sequoias in Vancouver's neighborhoods and on West Vancouver's hills that we all became aware of the fact that the Lower Mainland was changing.
      —Carole Taylor[65]

People are often thought of, and think of themselves, as being rooted in place and as deriving their identity from that rootedness. The roots in question here are not just any kind of roots; very often they are specifically arborescent in form.
      —Liisa Malkki[66]

One of the most bitter sites of contention between established homeowners and the recent Hong Kong arrivals was the removal of mature trees from the lawns and gardens of houses slated for demolition and rebuilding. Numerous community movements, such as the Kerrisdale-Granville Homeowners' Association (KGHA), arose in the 1980s to combat the "destruction" of the landscape and to protest the loss of ambience and character in the west-side neighborhoods of Vancouver.[67] In a neighborhood action against the felling of two giant sequoias in April 1990, for example, KGHA members protested by tying yellow ribbons around the trees a few days before they were scheduled to be removed. The protesters joined hands around the trees and left informational placards where passersby could read them. A month later, after the two sequoias had been felled, forty neighbors from the area planted new sequoias in a city park to commemorate the loss of the older trees. The community organization invited students from local schools to help plant the new trees. "An eight-year-old boy helped turn the soil," reported an article in the *Courier*. Cindy Spellman said of the planting, "It's worth it because we are planting the new trees for our children."[68]

The removal of mature trees and gardens from west-side neighborhoods caused more anger and resistance than had the demolition of apartment buildings or the construction of the so-called monster houses. Trees were linked with an image of Vancouver that was extremely important to many white residents. The tree-lined streetscape and even specific trees and gardens were identified with the essence of a west-side way of life; many west-side neighborhoods were established in the image of a pastoral Britain of the preindustrial era, and removal of the trees threatened this image.[69] The free-flowing landscaping and mature trees operated symbolically as a link to an imagined aristocratic past, but also as a bulwark against change, particularly pernicious change from the outside. Canetti writes of the forest: "Another, and no less important, aspect of the forest is its multiple immovability. Every single trunk is rooted in the ground and no menace from outside can move it. Its resistance is absolute; it does not give an inch. It can be felled, but not shifted. And thus the forest has become the symbol of the *army*, an army which has taken up a position, which does not flee in any circumstances, and which allows itself to be cut down to the last man before it gives a foot of ground."[70]

As the trees were also considered to be living beings, their destruction was perceived in a particularly serious light. Resistance took on the character of a moral crusade. Those who destroyed trees were depicted as morally wrong or depraved; at the least they were misguided or uneducated. One activist protested the felling of two sequoia trees at 6425 Marguerite in Kerrisdale by invoking religious language in her speech at a neighborhood protest: "These trees are part of the soul of the neighborhood."[71] After their removal, a reporter described the remaining empty space in terms of death and burial, again recalling metaphysical imagery: "All that remains of what one horticulturalist called 'the most perfectly matched sequoias in Vancouver' is two earth-covered stumps."[72]

Investing territory and things with human characteristics allows the naturalization of links between people and place. The connection between people and place, metaphorically expressed through tree imagery, is important not only in establishing, confirming, and romanticizing those who are "of the soil," who are part of a traceable, genealogical tradition, who "belong," but also in identifying and demonizing transients or sojourners arriving from elsewhere, those who are "without place." The condition of placelessness or rootlessness in society is perceived and represented as pathological and described so in moral terms.[73]

The moral element associated with these disturbance was linked to a positive depiction of "the country" and a negative image of "the city."[74] As a dichotomy between city and country morals and ways of living emerged, the Hong Kong Chinese immigrants often became identified as essential city dwellers. These cosmopolitan urbanites, carriers of the ill effects of accelerated modernism, were set up in opposition to and in confrontation with those who "love gardens." This became clear when I interviewed a west-side homeowner in September 1990.

V: What I feel personally, I'm European, so I very much like the garden. Our garden stretches quite a ways so that it's actually balanced with the house. You know how the houses in Europe are in proportion with the garden in general. You look at Spain and France and Germany, it's been the same for years. Although people have come and added to it, it's kept that balance. That's my difficulty. I love to see the houses in proportion with the garden. The problem is, in Hong Kong they have very little land. Here they come and there's so much land and so what do they do? You see?

K: Is it changing so quickly that you have a sense that it's going to be different for your children?

V: Well, my other area that I lived in between Oak and Forty-first and Forty-ninth, you wouldn't recognize it, it's completely changed. All the houses are torn down, and they were a right good size. But now they're huge, immense houses with very little garden. . . . Because I really wonder if they love gardens.

In this statement, a European heritage and love for a well-proportioned garden are set up in contrast with the Hong Kong tradition, in which an inherent love of the garden is dubious. The garden conjures up images of large wooded estates, authenticity, rootedness, tradition, folk, soil, a past.[75] The contrary image, of urban space and the lack of a proper "balance" of house and garden, distills meanings of rootlessness, sojourn and transience.[76] In specific neighborhood actions against tree removal, these images were invoked time and again, with the protection of mature trees coded as the protection of tradition, heritage, and a nostalgic, communally remembered past.

Both nation and culture are conceived in these rooted, territorialized, and essentializing images and terms. Transience or "decontextualized" culture threatens to denature and spoil these images of self-identification.[77] Uprootedness is strenuously resisted, particularly by those who already feel a threat to personal or national identity. The loss of an arborescent, genealogically traceable connection to the past is connected with the loss of an imagined future community.[78] The symbolism of using an eight-year-old to "turn the soil" reflects a perceived movement of time from the heritage and traditions of the past to the children and community of the future. In this vision, those who control the past control the future.

Apart from the sense of continuity that is advanced in these types of actions, there is also an implicit connotation of the right to judge what are appropriate and inappropriate activities on the landscape. The right to participate in the production of the landscape and its associated symbolic meanings is held only by certain populations. These populations, identified as "rooted," maintain and reflect the "correct" sensibilities appropriate to the land. Historically, the ability to participate in landscape production, and hence in society itself, has been predicated on highly racialized and gendered grounds.[79]

Graphic images that connected the preservation of Vancouver's trees with the preservation of Vancouver's future were evident in brochures such as the January 1991 public information announcement concerning the proposal of new tree bylaws in the city. In this graphic, the white structure of the tree closely resembled that of a pregnant woman. Inside the pamphlet was an image of a tree composed of roots or branches dividing into smaller and smaller limbs. The image resonated as one of connectedness and of natural and organic growth and fecundity through time.[80]

In contrast, sojourners and city dwellers, who lacked the direct connection to the past and the soil, were depicted as unable to participate in or even understand the imperative to guide and manipulate the future. As they were rootless (cut off), they could be portrayed as fundamentally unsolicitous of the needs of future generations. Although the majority of those who resisted tree removal considered and presented their struggle as one of environmental protection, the aforementioned representations were vivid enough for many developers to contest the antitree removal movements as racist. When asked if his organization's resistance to the felling of the two sequoias in April might be construed as racist, John Simmons said in a newspaper interview: "I don't care who is doing it. If people are doing something that is destructive to the neighborhood, there is going to be a reaction."[81]

Despite Simmons's stance, however, the location of the two sequoias on the property of a Hong Kong Chinese immigrant and developer, whose daughter was a highly prominent real-estate broker doing extensive business with Hong Kong, gave the conflict a racial cast. In many of the newspaper articles on the sequoia cutting and protest actions, the daughter's name was mentioned in connection with the event. During a period of extensive media coverage, her Mercedes was vandalized. In Hong Kong, the story was a controversial high-profile news item for more than a week, with the clear perception overseas that the resistance protests were racially motivated.[82]

More than thirty letters concerning tree management were sent to the Vancouver Department of Urban Planning between

1988 and 1990. Most came from west-side neighborhoods and indicated a desire for more tree regulation by the city.[83] On March 7, 1989, city officials petitioned the province for the authority "to regulate the destruction or removal of trees and for making different regulations for different areas of the city."[84] Although the B.C. provincial government declined to give the city authority over tree cutting on private lots, city officials initiated bylaws in early 1991 designed to protect mature trees in RS-1 (single-family) neighborhoods. These bylaws drew from a 1990 study by a consultant team of planners, landscape architects, municipal lawyers, and arboriculturalists, "Trees on Single Family Lots: A Program for the Protection of Trees on Private Property."

In this planning guide, "heritage trees" were outlined by neighborhood and by individual tree. In a complex mathematical calculation, trees were valued based on their diameter, species type, condition, and location. With the value of trees determined in this supposedly objective manner, the tree debate was removed from the emotionally charged atmosphere of racism. "Rational" planning instruments and scientific criteria calculated the value of individual trees in a separate sphere, away from the messy contamination of ideological debate. The introduction of racism as an element of the debate could then be characterized as an example of nonreason and consequently dismissed.

The Hong Kong Chinese immigrants contested both the bracketing of racism in the private sphere and the communitarian rhetoric of authenticity and belonging that accompanied the tree debates. For example, disrupting the hegemonic narrative of nations and roots, a group of Chinese executives contested the urban and rootless images of the Hong Kong Chinese with a business fair, "TRANS/PLANTS: New Canadian Entrepreneurs." The fair, organized by the Hong Kong–Canada Business Association (HKCBA), featured immigrant entrepreneurs engaged in trading and manufacturing. The intent of the exhibition was to counter the image of Hong Kong Chinese investors as exploitative real-estate speculators by demonstrating the many tangible economic contributions of the Chinese entrepreneurs in areas of the productive economy. Real-estate speculation, or "flipping," was a source of great anger for long-term residents, who felt that Hong

Kong buyers were interested in houses only for profit, rather than for establishing "roots" in a long-term home. I interviewed the president of the HKCBA in March 1991 in Vancouver.

> B: Because of the newspapers' attacks on Orientals mainly investing in real estate, we came out with an exhibition. We didn't invite anybody that's involved in the real-estate business. We invited only those people doing manufacturing and trading to participate in this exhibition.
>
> K: Was it a deliberate attempt to counter the ideology that the media has been presenting?
>
> B: Yes, it's a deliberate attempt put out by our organization to counter that situation. To show to the public that there is a group that is doing non–real estate.

The term "trans/plant," or "transplantation," elicits images of displacement in a positive sense, as something that remains viable and alive in a new setting. Rather than uprootedness, which resonates as an unwanted disrupture, it evokes a degree of will or power in the movement. Malkki writes of the term "transplantation": "It strongly suggests, for example, the colonial and postcolonial, usually privileged, category of 'expatriates' who pick up their roots in an orderly manner from the 'mother country,' the originative culture-bed, and set about their 'acclimatization' in the 'foreign environment' or on 'foreign soil'—again, in an orderly manner."[85] Using the image of the transplant as a Chinese entrepreneur, rather than a white colonial expatriate, again reverses established meanings of nature and normality. "Civilization" is brought from the seedbed of Chinese entrepreneurial capitalism and deposited to reflower on Vancouver's fertile but largely untilled suburban soil. Rather than the pastoral Eden of a preindustrial countryside, "the originative" site is Hong Kong, the original man is Chinese, and the origins of civilization are urban capitalism.

In this case, the reworking of a roots narrative by Hong Kong businessmen disrupted the normative cultural associations of British Canadian liberalism as they had been established in Vancouver. At the same time, the new narrative functioned as a smooth supplement to an ideology of free trade and free markets. The counter narrative to the roots metaphor was associated with a mobile sense of transplantation. Thus the new spaces of the entrepreneurial transplant dovetailed well with the neoliberal spaces of global capital accumulation.

## CONCLUSION

Liberalism connotes a spaceless plain, and as a result, no *explicit* connection between territory and political belonging is generally acknowledged in British liberal theory.[86] Yet despite its denial of this connection in the abstract, those who borrow from this tradition *assume* a culturally inscribed political territory based on the normative understandings of British culture. In white westside neighborhoods of Vancouver, for example, long-term residents relied on the historical and geographical contingencies of spatial and cultural privilege in the urban environment to delimit universal equal membership in the political community.

With the arrival of wealthy Chinese immigrants from Hong Kong, however, the implicit, British-oriented, cultural assumptions relating to the urban landscape came to light, and the liberal discourse of universal urban entitlement and political membership was exposed as a false promise. More than this, the delimitations to equal membership were shown as the result, not of accidents of poor planning or individual discrimination, but of exclusions *immanent* in systems of land governance. Exclusion by race and by class was integral to the ways profit was made from real-estate development in Vancouver historically and, as the recent immigrants demonstrated, was directly implicated in the language of rationality, order, authenticity, and normality in the discourse of urban planning.

Drawing on variations of Canadian liberalism, as well as the contemporary discourse of neoliberalism, to bolster their opposition, transnational migrants disturbed the practices of land governance in the city by refusing to accede to the distinctions and cultural norms of the Vancouver landscape. Further, they reworked the signs and symbols of distinction, authenticity, and national belonging, such as trees and roots, to privilege a neoliberal discourse of global space and a Lockean interpretation of a nature that becomes valuable when labored on for profit. In their challenge, the issue of race was a central, if often unspoken, factor that operated as a condensation point for discussions around belonging, individual rights, and the real or perceived abrogation of those rights.

Despite their own historic involvement in land governance, members of the property industry used the issue of race more out-spokenly to discredit certain kinds of state intervention in the landscape, and to promote a neoliberal vision of a less divisive form of governance based on the image of a fundamentally ahis-torical and ageographical community. The conflicts between the Chinese immigrants and the white residents, and the use of nationally oriented symbols and narratives of belonging and authenticity by some white activists, provided the rationale for these developers and neoliberal politicians to attack state policies that could be construed as interfering in the "fairer," race-blind workings of the global market. In their discourse, the open and accessible plain of free-market globalization was juxtaposed with the exclusionary spaces of the national community. In the highly racialized struggles over the control of urban land in Vancouver, they posited neoliberal policies as producing racially untainted, universally accessible space, in contrast with the false promise of a liberalism based on exclusionary British cultural norms.

# 5   Domesticity, Race, and Uncanny Homes

IN VANCOUVER, the demolition of smaller homes on high-priced lots and the construction of large "monster" houses became associated with the influx of wealthy Chinese transnationals, many of whom purchased these buildings in the 1980s and 1990s. The construction of the so-called monster houses, primarily in white west-side neighborhoods, was the focus of heated struggles for over a decade. What exactly did these homely spaces mean? How were they produced historically and sedimented spatially in the imaginations of residents? What were the broader notions of domesticity, class, race, and nation in the old homes and neighborhoods that were rendered unstable by the entry of the global, entrepreneurial, and raced bodies of the Chinese immigrants? And how were liberal narratives of space, modernity, and political membership disturbed by this movement?

As with most urban milieus in the West, middle-class imaginings of the proper domestic life and lifestyle in Canada developed historically and spatially in close association with the growth of capitalism and liberalism. Modern collective memories of house and home, however, tended to elide these origins, especially the class- and race-based systems through which these processes developed. For example, through its association of the normal and the rational with specific types of cultural distinctions and spatial practices—especially, as in the cases discussed here, those associated with the house and garden—liberal thought helped naturalize racial distinctions and conflate these with cultural differences.

Long-term liberal understandings of the normal and the rational, however, were disturbed in the domestic sphere of Vancouver life as a result of the different transnational cultural practices and imaginings of the recent Hong Kong Chinese migrants. Cultural differences became apparent in many areas: beliefs around geomancy or feng shui, the importance of size and new-ness rather than architectural or landscape aesthetics, views on nature and gardens, assumptions about appropriate family rela-tions and size, and conceptions related to the meaning of the home. These cultural differences were the basis for antagonism between older and newer residents; they also served as a foil for those few Vancouverites who believed that Vancouver and Can-ada should remain predominantly white. This assortment of white supremacists and ordinary racists consistently and delib-erately conflated racialized conflict with the supposedly "deeper conflicts between cultures and civilizations,"[1] which had the double effect of diminishing the power of antiracist language and augmenting the position of neoliberal strategists, who sought to link all resistance to urban and cultural change with racism.

As a symbolic space of the nation, the home became a key icon in cultural and racial conflicts, and in the struggle to conflate the two. The home in the West stands not only as metonym for the nation, but also for a set of ideas associated with domesticity and the natural order of human relations.[2] These ideas are generally employed in relation to gender but contain implicit meanings that are profoundly racialized and link histories of colonial empire to contemporary struggles over space and design. Historians of domesticity link the emergence of the domestic as an ideological construct with the rise of industrial capitalism in Europe in the seventeenth century.[3] With the expansion of market society came a host of related processes, including a gendered division of labor that rested on an increasingly stark separation of public and pri-vate realms. As Comaroff and Comaroff note, it has become a commonplace to observe that "the emergence of a developed "domestic domain"—associated with women, unwaged house-work, child raising, and the "private"—was a corollary of indus-trial capitalism." They also remind us that domesticity was "inte-gral to the cult of "modernity" at the core of bourgeois ideology."[4]

The relationship between domesticity, industrial capitalism, modernity, and the separation of public and private domains is also a central theme in many scholarly accounts of the spatial technologies of power wielded in various *colonial* endeavors around the globe. These accounts associate domesticity with the "modern" form of family life, that is, with the rational ordering of the spaces of the home and the division of labor within and outside the home. When transported to the colonies, this construct of modernity served as one of many spatial techniques of conquest and control. For example, in his study of the U.S. colonization of the Philippines at the turn of the century, Rafael showed how the modernizing imperatives of domesticity operated to initiate and enforce new kinds of bodily values, spaces, and behaviors for both U.S. and Philippina women, as well as to alter the physical architecture and landscape of the home.[5] Thus ideas of domesticity were expressed through the rituals and routines of everyday life and linked with the modern and the rational. In the process they facilitated the entry of market economies and served in the colonizing projects of many nineteenth- and early twentieth-century empires.

In addition to creating a domestic moral order that transfigured and disciplined gender relationships, the discourse of domesticity in colonial modernity assumed a racialized separation of spaces and behaviors.[6] White hegemony in the overseas colonies was predicated on the separation and disciplining of the "native" body; this was achieved by reordering space on a number of scales. The impact of this reordering on subjectivity formation and the experience of time and space was immense. As Mitchell writes: "Colonising refers not simply to the establishing of a European presence but also to the spread of a political order that inscribes in the social world a new conception of space, new forms of personhood, and a new means of manufacturing the experience of the real."[7] The spread of this new political order and the new experiences of "the real" were frequently based on the dissemination of highly racialized conceptions of the appropriate spatial positioning of white (British) bodies and nonwhite Others.

What I want to examine in the context of late twentieth-century Vancouver is a somewhat different scenario. As noted by Comaroff and Comaroff, the colonial endeavor influenced the construction of modern domesticity both in the colonies overseas and *at home*.[8] The dissemination of Western modernity was never one-way, nor was it ever a homogeneous process; it was shaped by the historical and geographical contexts of specific colonial encounters, and these encounters operated back on the metropoles themselves. A "British" construct of the domestic in Canada as it was formed and sedimented through time was inflected with Britain's colonial history, as well as with the transposition of these meanings to British Columbia. How then shall we think about domesticity as it plays out in the spatial struggles of late twentieth-century Vancouver?

Two British national narratives pertain to my study here: the equation of Britain with empire, and the equation of Britishness with whiteness. Both are well documented in the academic literature, with the latter explored most convincingly in Paul Gilroy's study *There Ain't No Black in the Union Jack.* The hegemonic formation that equates Britain, empire, and whiteness remains deeply ingrained in popular white middle-class consciousness in England, as evident in the reaction to both the Falklands "crisis" and the Rushdie "affair."[9] The perceived threats to this "structure of feeling," especially with respect to immigration from outside Europe, are periodically manipulated by politicians such as Enoch Powell in his now infamous speeches of 1968 and 1971, as well as by more contemporary politicians such as William Hague.[10]

Canada's relationship with Britain is complex, but since early trade and settlement times, the two countries, aside from Québec, have maintained strong political, economic, and cultural ties.[11] The cultural ties remain strongest, perhaps, in British Columbia, a province that has consistently drawn a high percentage of immigrants from Britain.[12] Many residents of British Columbia claim a strong sense of cultural identification with Britain, including, for some, a belief that the heritage and legacy of the province is and should remain primarily "white."[13] This identification was seemingly confirmed with the relatively large success of the far-right Reform Party of British Columbia in the 1990s, which campaigned largely on an anti-immigration platform. As with many such agendas, the implicit text of the party concerned white heritage and the necessity to protect the British character and legacy of the province.

White hegemony in British Columbia was first spatially secured with the establishment of government reserves for native tribes in the province. Smallpox and other epidemics, which had arrived with earlier traders, had effectively exterminated a majority of the aboriginal peoples through the preceding decades.[14] The "reserves" policy established an even more complete segregation between the natives and the settlers than most colonial encounters. However, the racial coding and spatial separation of the native tribes was just one aspect of a much broader system of conceptual control over the status of all settlers, immigrants, and natives defined as nonwhite by the legislative assembly and the provincial parliament of British Columbia.[15] This racial coding, for example, quickly positioned Chinese immigrants as perennial outsiders, the Others against whom British Europeanness could be measured. As Anderson wrote of this segregation:

> In British Columbia, as in other British outposts, we shall see that a colonial bureaucracy quickly specified, officially, the characteristics thought to differentiate outsiders from those deemed to be legitimate citizens. In doing so for the "Chinese," it will be argued, the agents of the Canadian state sanctioned a concept behind which lay the most divisive world view of "us" and "them." In the ambition to build a dominant "Anglo" identity and community, the state sought to secure popular legitimacy by defining people of Chinese origin in opposition to all that could be made to stand for "white" Canada. With all the backing of the imperial mission, politicians took as their mandate the making of a European society in all its institutional and private domains.[16]

This power to define by racial category was paramount in setting up the finely tuned spatial controls that demarcated white British settlers from all others and established the normative associations of particular spaces with particular groups. Soon after the period in which Britain established its spatial presence in China through "gunboat diplomacy" and the colonization of Hong Kong, British settlers and the incipient state apparatus adopted similar imperial strategies in British Columbia. Using racial ideologies that defined Chinese immigrants and other nonwhite groups as outsiders, bureaucrats, politicians, judges, capitalists, and other leaders of society shaped the social milieus in which the immigrants arrived, as well as the literal spaces in which they were allowed to move and reside. Through time, as Anderson remarked, the reciprocal relationship between social definitions and spatial production became mutually reaffirming; the normative association of specific groups of people with specific kinds of spaces and homes was firmly established and sedimented: "State practices institutionalized the concept of a Chinese race, but it was in space that the concept became materially cemented and naturalized in everyday life."[17]

More than any other immigrant group, the Chinese were controlled and disciplined by immigration regulations, taxes, real-estate regulations, and other kinds of spatial restraints, all of which drew on and reconfirmed racial definitions that were part of a broader system of imperial European hegemony.[18] The Chinese body was inflected with multiple negative attributes, including unsanitariness, addiction, lustfulness, and disease, and the Chinese as a social group were positioned as gamblers and wastrels, people who enjoyed living in overcrowded squalid housing.[19] These representations were then used to legitimate the strict confines in which people of Chinese descent were allowed to live and work.

Although the Chinese became enfranchised in Canada in 1947 and immigration laws were liberalized in 1967, the hegemonic definitions and representations of earlier years were spatially sedimented through time, and informal covenants restricting home ownership for nonwhite buyers continued to operate in many neighborhoods throughout the following decades.[20] Those privileged by these spatial norms aided in the naturalization of the imagery of these primarily white neighborhoods with a particular way of life. The imperial legacy of these racialized spaces receded from contemporary memory yet continued to operate in all kinds of subtle and not-so-subtle ways.

In this context domesticity, as a particular order of dispositions and meanings, was constituted by the deeply layered spatial hegemony of race construction in British Columbia. The meaning of home, the appropriateness of certain kinds of bodies in certain kinds of spaces, the relationship of these bodies to the market economy—all were bound up with the temporal and spatial relationship between white British settlers and Chinese immigrants in Canada.[21]

The rupture of this relationship occurs through the inversion of normative assumptions about homeownership, belonging, empire, and race; it is fought out in spatial terms vis-à-vis the images of landscape, architectural design, and the embodied spaces of the home. In what follows I show how what constitutes liberal Canadian understanding of the domestic—especially with respect to normative concepts of public and private, normal and abnormal, rational and irrational, and white/non-white—is disrupted when Chinese entrepreneurs move into the monster houses of Vancouver's west-side neighborhoods.

## RACING THE MONSTER

I grew up in Shaughnessy, on Balfour Street, and have watched closely the changes happening within it. I am saddened and disgusted when I walk through it today to see so many of the trees and houses gone, only to be replaced by hideous monster houses. . . . I talked to a construction worker who was working on one of these new atrocities they call a house; . . . he said, and I quote, "The house is a piece of shit, and will probably be falling to pieces in ten years." So, is this what Shaughnessy is to become? We need assurances that the character of the neighborhood will be maintained!
        —Letter to Vancouver City Council, April 10, 1990

There are three basic forms of the generic monster house, which correspond with the zoning regulations before 1986, and to the zoning revisions of April 1986 and April 1988. Features that most of the houses share include the following: a rectangular, usually symmetrical shape that occupies the maximum allowed lot-surface coverage and height, and built to the maximum allowed square footage in size;[22] a large entranceway with double doors and a two-story entrance hall; large, symmetrical unshuttered windows with occasional glass brick detailing; and external finishes generally composed of brick or stone, with some stucco, vinyl, or cedar siding (see Illustrations 1 through 3). The interiors were spacious and plain, with an average of five to seven bedrooms and an equivalent number of bathrooms.

The style was a great source of controversy, especially in Shaughnessy Heights, and the sheer scale of the buildings when juxtaposed with the older houses on the street was also a major sore point for long-term residents (see Illustration 4). Illustrations 5 and 6 demonstrate the contrast in scale. The "monster" house in Illustration 6 was constructed at Fifty-fourth and Wiltshire in Kerrisdale, on the same block as the two houses in Illustration 5.

Until 1986, monster houses often included interior garages, but garage space was incorporated into floor-space ratio (vis-à-vis its lot size) with the zoning changes of 1986, which made interior garages less attractive to developers trying to capitalize on

ILLUSTRATION 1. Monster house with two-story windows and double doors. All photographs by the author.

ILLUSTRATION 2. Monster house with mixed brick and wood siding.

ILLUSTRATION 3. Monster house with concrete exterior, glass brick, and large gate.

ILLUSTRATION 4. Old and new house styles juxtaposed.

ILLUSTRATION 5. Small bungalows in Vancouver.

ILLUSTRATION 6. Newly built house on same block as bungalows shown in Illustration 5.

space. Until 1988, two- or three-car garages were often built in front of the house. The zoning amendment of 1988 disallowed street access for cars, however, and thereafter the garages were placed in back.[23] In the effort to reduce the sheer bulky envelope of some of the new houses, especially in contrast with older houses on a block, one zoning measure had the unfortunate effect of encouraging developers to try to retain more indoor space by producing asymmetrical rooflines and windows (see Illustration 7). The houses were usually landscaped with a fence or stylized hedge surrounding the lot, which was extremely uncommon in Vancouver at that time. In general, the landscaping was more formal and minimalist than traditional yards, and occasionally the entire yard was paved (see Illustration 8).

The monster houses built during these years thus varied in form and style depending on the zoning laws in effect at the time of their construction. However, when I asked long-term residents to describe the monster houses, they often responded with some irritation and surprise, "Well, haven't you seen them?" The generalized image of the monster house was so prevalent in the popular imagination of the 1980s that any *particular* description

ILLUSTRATION 7. Asymmetrical rooflines and uneven massing.

Illustration 8. New house with paved yard and minimalist greenery.

seemed somehow less than warranted. Annie Humphreys, a leader in Concerned Citizens for Affordable Housing (CCAH), drove me around her Kerrisdale neighborhood for more than three hours, pointing out the new homes and the numerous recent changes to the area. Her conversation wove together the abstract and the concrete, memory and desire. From a sharp comment on an ugly architectural feature and poor quality materials she shifted to nostalgia for the garden that had been displaced:

> This building up here, which is locally known as "The Fortress," has iron bars on every single window and every single doorway [1200 block of Forty-eighth Street]. If you go up the side of this smaller house alongside this one, you go through a tunnel, virtually, because they've overbuilt their fence. . . . This is a new one on the corner. It just goes on and on, and it's so ugly because you get these columns at the front, these Greek columns which are two stories high. And it's a disaster [Montgomery and Forty-ninth]. These three houses in a row—there used to be such beautiful gardens here. People prided themselves in their lawns and garden area. . . . Look at this one. It's gone. And they haven't done anything about that garden [Montgomery and Fiftieth]. This one is a prizewinner—well, I'm sure that that house would look very beautiful on acreage appropriate for the size of it, and it would be

wonderful in California because you've got your palm tree that grows up that's two stories there in that front window. But here, amongst these smaller houses... This whole block, ... this is where that prizewinning garden was. And this house here has the most rhododendrons I've ever seen in my life. And you know that that house is doomed.[24]

As Annie led me through her neighborhood, an acute sense of loss and personal distress permeated her descriptions of the new houses. In 1988, the house next to hers had been demolished and a much larger one built in its place. During its construction, the effluent from the building materials ran into her lawn, workers left garbage and building waste on her property, and the noise was horrendous. The new building was so much bigger than hers, it blocked the sunlight, and she felt that the new neighbors could look into her backyard from their second-story windows. She told me that she had written to City Hall several times, describing the ubiquity of bulldozers and the unpleasant changes in the neighborhood, but that nothing had changed.

Annie's anger about the monster houses and the demolition of low-rise apartment buildings in Kerrisdale spurred her to become an active member of the CCAH.[25] She was frequently interviewed by the news media, and became a familiar spokesperson for the community. When I visited her in 1990, she gave me three large scrapbooks and several shoeboxes full of newspaper clippings about monster houses, apartment demolitions, and general neighborhood change. She had also collected articles on immigration, education, population statistics, and the Hong Kong Chinese. Several members of other residential organizations had written her for advice about community organizing or to request or offer support; their letters were in the files.

When I examined Annie's collection, a letter and a brochure from white supremacist organizations caught my eye. The brochure was from an organization called the British/European Immigration Aid Foundation and was headed with the saying, "The *True* North Strong and Free." The text began:

This Foundation is based on and will build on patriotism. We need to be patriots. Deep down, we believe most of us are quietly proud of our British and European roots. We treasure more than we realize the common traditions, the shared values which form the very basis of

our Canadian nationhood, the strong and abiding links of kinship. In
the present immigration picture, the importance of these are being
swept aside. For Canada's sake, they must be reaffirmed. . . . Patriotism, inspiration, and united determination will do it. PRESERVE
CANADA'S HERITAGE.

At the bottom of this brochure, Annie had written a telephone
number for the organization and a meeting place at the Kitsilano
Recreation Centre for June 26, 1989.

The letter was from a white preservationist group, the Spokes-
man. This prowhite group was more vociferous in its demands
for an end to "multi-racism" and the immigration of nonwhites.
The writers made it clear that the influx of other "racial groups"
signaled the deterioration of "our way of living." At the bottom
of this more virulently racist document Annie had written,
"strange people!!"

Annie's response to these unsolicited letters and advertisements from white supremacist groups positioned her on a complex spectrum of racially based reactions to the changes in her
neighborhood. For Annie, the aim to preserve a so-called Canadian heritage was significant enough to attend or think of attending a meeting that emphasized the curtailment of non-European
immigration. At the same time, she found it strange when a group
with the same basic agenda but using much stronger language
apprised her of their organization's intents. Annie's feelings about
the arrival of the Hong Kong Chinese were mixed. Although she
voiced sympathy with "the people that are coming from the Asian
countries and feel very threatened by the communist system,"
she also felt that their capital should not be used in her neighborhood. "Of course they wanted to get their dollars out. But let's
get them to put their dollars where it ultimately benefits the
country. And I can't see how people getting evicted from their
homes is benefiting the country."

In her interviews with me, Annie positioned herself as a community organizer and defender of the rights and values of the
white upper-middle class women being evicted from their Kerrisdale walk-up apartments. As a divorced professional musician,
she identified with the plight of the elderly women whose apartments were being demolished to make way for twelve-story luxury condominiums. She portrayed her community activism

against monster houses and unwanted demolitions as her primary role in CCAH. Her more private feelings about the underlying cause of these occurrences—a government sell-out to Hong Kong investors and developers—were much more opaque. Annie was wary of being labeled a racist and attempted to separate her work as an urban activist against unwanted economic activity from what she considered the "social" issue of racism.

> It isn't that we're against the people coming from Hong Kong, but . . . this is not a social issue. They're trying to make it into a social issue and they point fingers at anybody who complains about this investment, and they say you're racist. Racism is a social issue, it is not a financial issue. This is two different arguments we're dealing with here. We're talking about *our* country not having the control, or not willing to have the control to put *our* population, the *Canadians*, in control of their own land. They are bartering our land in order to get all those billions of dollars from another country in investment to stimulate the economy . . . and of course it makes the government look glowing in their balance sheet. And the city.

Annie believed her views were not racist because racism is "a social issue" and what she objected to was financial, the sell-out of Canada's heritage for money. Her equation of Canada's heritage with British culture and white people was so natural, in her mind, that she could not understand the implications of maintaining that heritage by keeping Chinese immigrants from investing in land and houses in the city. Her sincere, if misguided, efforts to separate the workings of capitalism and racism, however, were constantly thwarted by those to whom the interlocking of the two processes was advantageous. For some groups, "racing" capital, projecting it, and maintaining it in the image of the Other furthers their own ideological purposes. Although Annie was careful to verbally position herself as an urban activist, her well-publicized role in the confrontation over the bulldozing of several apartment buildings in Kerrisdale caused some individuals and organizations to position her in a different manner. The white supremacist groups and individuals who sent her unsolicited mail linked her actions to protect the neighborhood and community from unwanted capitalist development with their own aims of an exclusionary immigration policy and the fomentation of anti-Chinese sentiment.

In her interviews with me, Annie made a point of declaring her stand to be one of opposing international capitalism and the negative effects that global real-estate investment were having on her community. She did not want her struggle to be tarnished or undermined by being labeled racist, regardless of her own personal feelings or dilemmas. She was not, however, able to deflect the desires of marginal groups, who perceived her as the perfect conduit for their messages of white supremacy.

The struggle over the foregrounding of race in these neighborhood movements demonstrates the crucial role that race construction plays in the contemporary neoliberal articulation of global capitals and international investment. Neighborhood activists' desire to keep race out of the monster-house controversy was met with the equally fierce desire of marginal groups to advance their causes in this public domain. And because of this activity, businesspeople and state representatives who benefited from international investment and real-estate development were able to point to the visible racism of these latter groups and proclaim *all* neighborhood slow-growth movements as fundamentally racist in nature.[26]

## DOMESTICITY AND THE ARCHITECTURE OF BELONGING

The *domestic* implies spatial arrangements in which certain practices of reproduction (children as well as certain modes of production) are situated. As a primary site at which modernity is manufactured and made manifest, the domestic serves as a regulative norm that refigures conceptions of the family from a largely temporal organization of kinship into a spatially manifest entity. The domestic with all its material and metaphysical accoutrements bridges the distance between seemingly public issues and the private concerns of families.

    —Rosemary Marangoly George, "Recycling"[27]

Aside from a discourse of loss that permeated most discussions of the monster houses and the demolitions of older buildings, a major concern expressed by white Kerrisdale and Shaughnessy residents related to architectural aesthetics. Without identifying specific features, older residents commented on the overall ugliness and inappropriateness of the new houses. Many felt that the homes were in bad taste, particularly for west-side neighborhoods. A *Vancouver Sun* editorial, "Monster Houses: A Matter of Taste," commented on April 2, 1990: "It's been said that some people's taste is all in their mouths. Those words must often come to the lips of passersby who behold the more bizarre examples of the monster houses inflicted on Vancouver. The neighbors use much stronger language." Innumerable statements in the press, letters to City Council, and research reports reflected aesthetic disapproval. The most common remarks can be paraphrased: "They're lot-line monsters"; "I just don't like them, they're ugly"; "They're okay, but not in this neighborhood"; "Ugly, box-like monsters"; "You know, you've seen them, they just don't fit in."

One oft-told anecdote was of the doctor's wife who spray painted, "This Is An Ugly Home" on one of the new houses in her neighborhood. Another, more esoteric monster-house graffiti asked in large, black, spray-painted letters, "Genius Loc?"[28] "Genius loci?" is a Latin phrase which asks: Is the guardian spirit of the place in its rightful context? The phrase operates on a number of other levels as well. "Genius" from the Greek refers to birth, and in this context could be seen to question the nativity or indigenous nature of the monster house. The use of "genius" could also function as an ironic play on the English usage, wherein the architectural genius of the project is questioned. Finally, "genius" from the Greek is often employed in a metaphysical sense and here could be seen as a subtle attack on the more material uses of the house by Hong Kong buyers as a site of profit rather than as a place of spirituality.[29]

The power of authority expressed and wielded by these graffiti writers is explicit. In the first example, the tacit assumption of the graffitist that her message of disdain would be acceptable to the white community is evident in the public nature of the act. Those who told me the anecdote were well aware of the woman's identity and related the story with some delight at the presumptuousness of her message and action. The smug assurance that the taste of other community residents was in accord with her own allowed the graffitist to express her opinion as a universal one. When she sprayed, "This is an ugly home" on the house, it was an act of mastery and control. In the second example, the arcane nature of the phrase positioned the graffito at the level of

"high" culture and highbrow taste. The elite knowledge necessary to unpack the meanings of the graffito was assumed as the
norm for the other residents of the neighborhood in which it was
displayed. Here the assumption of a shared esoteric taste allowed
an in-joke to be spread among community residents, yet implicitly excluded outsiders whose knowledge of Greek, Latin, and
the literary canon was inadequate for an understanding of the
phrase's many-layered references.

Several wider interpretations can be offered here. Many scholars have made the link between concepts of domesticity and
empire, tracing the connections between the establishment,
maintenance, ideology, and spatial layout of the home (both the
home "at home" and the colonial home) and the maintenance of
colonial rule.[30] The home, as the primary repository of the domestic, functions as an important ideological referent in narratives
of empire. But this correspondence is shattered in at least two
ways by the movement of wealthy Chinese immigrants into
wealthy white neighborhoods in Vancouver. First, there is the
obvious *inversion* of empire, in that nonwhite members of a former colony purchase the formerly white-controlled house itself,
the legacy of colonial settlement from Britain. Second, the unequal
economic relationship of colonizer/colonized, which has historically underpinned white middle-class conceptions of domesticity,
and which was often legitimized on the basis of racial narratives,
is also inverted.

With regard to the inversion of empire, it is interesting that it
was a white *woman* who spray painted the monster house, and
that many of the housing demonstrations and community movements discussed in the preceding chapters were led by women.
Traditionally, women have been depicted as the defenders of the
hearth, as fundamentally conserving of cultural dictums and
norms, and conservative with respect to changes in the traditional religious and sociocultural landscape. Rosemary Marangoly
George argues that British women in the colonies (she uses examples from India) gained an "authoritative self" through their management and control of the colonial home. As a result of the general assumption that the skills of household management were
necessary for running a successful empire, the colonial woman
was seen as an important buttress to empire. The English woman's

greater status of political and personal authority was thus won through her performance in upholding the imperial project. It was, furthermore, a project that enabled her to cross and blur the distinctions of public and private, as George points out: "This authoritative self was defined against a racial Other in encounters that were located in space that was paradoxically domestic as well as public: the English home in the colonies."[31]

Did the upper-class white woman of Kerrisdale feel a threat to an "authoritative self" historically won and spatially maintained via the effective management and control of the colonial home? Was the attack on the monster house an attempt to preserve a traditional British colonial landscape, with all its multiple associations? In this regard it is useful to remember that Chinese "houseboys" did much of the housework of middle- and upper-class white society in early twentieth-century Vancouver. The white woman's struggle for domestic reform and a reworked hierarchical status related to domesticity and housework in the mid to late twentieth century challenged public and private divides, but it operated on a domestic logic and terrain different from that of the struggle from Chinese houseboy to Chinese homeowner. The disruptions of public and private and the constant tactical redrawing of boundaries, usually examined in light of gender relations, may be read quite differently through the lens of race relations. The struggle over the meanings of domesticity involves multiple actors and histories, any one of which might capitalize on and feed off the others at a given moment in time. This different kind of inversion of normative relations and understandings also serves well as an epistemological reminder, as Hansen notes, that "the domestic is not everywhere nor exclusively organized by gender, but also by class and race relations."[32]

White residents of west-side neighborhoods often used local elements and dynamics of spatial and architectural change to describe the unwelcome transformations of their communities. Rather than depicting a process of change that included social and economic ramifications, such as the introduction of Chinese people into the neighborhood, many residents focused their criticisms on relatively neutral problems, such as the pollution caused by the new buildings or their poor architectural design, which led to a lack of sociability. Some pointed to the high level of waste and problems of disposal of the razed building materials; others referred to ecological concerns related to tree and shrubbery removal; many discussed a fear of social breakdown occasioned by the new "un-neighborly" architectural styles. The monster house, in this usage, functions as what Spivak calls a "screen allegory," where broader narratives are occluded through the concentration on a local, restricted plot.[33] The expressed fear of waste, couched in the environmental lingo of pollution and disposal, serves as an alibi for the more general but unexpressed fear of the filth of the Other.

The unnecessary, polluting waste related to the demolition of habitable houses was perceived as an example of unsustainable development, of the unsustainability of the new way of life of the immigrants. Like the wider narratives of environmental unsustainability, catastrophe seemed to be just around the corner; the rhetoric functions as a warning signal that societal collapse may occur unless the processes causing this waste are immediately halted. The alien polluting body of the immigrant operates as a metonym for the pollution of the clean Canadian land and the decline of its nurturing and sustaining capacities. The production of waste is also a visible manifestation of the creative destruction of capitalism, whose inexorable mastication of the landscape as a result of rent-gap speculation is brought "home" to wealthy homeowners through the destruction of the house and land next door.

The architecture itself was also used as an alibi for lack of contact. Suzanne Barlow, a young homeowner and homemaker in Kerrisdale, felt that the placement of the garage in the new monster house next door was a crucial factor in the ensuing lack of contact with her new Chinese neighbors. She told me in a 1990 interview:

> These big houses that they're putting up with very little backyard. . . . It gives me a funny feeling, because it's like they're putting barriers up. The garages that would front here in the old style [she points to the backyard], if they had children, we would see them. But they get in their car, they open their garage door, and they go. And there's no mixing. When people cross your alley you usually run into them because you're getting in or out of the car or taking your garbage out, but you don't get any of that. There's a wall. And that's kind of the nice thing people usually like about neighborhoods that's lost now.

Although the spatial layout of the monster house, garage, and garden may indeed exacerbate social atomization, the idea that architectural design is responsible for social breakdown also operates like the screen formed by dream displacement discussed by Freud. In this kind of displacement, "acceptable representations" are created and substituted for "unacceptable wishes."[34] Placing the responsibility for the lack of contact and neighborliness on the monster house itself shifts attention away from more structural or institutional problems in social relations involving class or race. In this case, design features of the monster house are criticized for causing a lack of contact between neighbors. These negative design features can then be cited as the more neutral rationale behind neighborhood anger and subsequent mobilization to curb the transformation of the houses and neighborhood. The more fundamental problems relating to social and economic relations and histories can thus be safely elided.[35]

In a letter to the editor of the *Courier*, April 4, 1990, a resident of Shaughnessy Heights linked the monster houses with an abnormal family life—one that violated the prevailing custom of controlled family planning.

> I agree wholeheartedly with the sentiments of Shaughnessy home-owner Donald Tuck (Monster Mishmash, March 28 issue): "we're concerned with the nature of the neighborhood." If I am not being too naive on the subject of "monster houses," I ask: Who, in this day and age of family planning, vigilant birth control methods, and the "norm" of 2.5 children per household, could possibly make use of houses containing seven-nine bedrooms and seven sets of plumbing plus two kitchens.... That is, if they are being used as "single family dwellings"?

The concept of the "normal" Canadian family is crucial here, as those occupying the new monster houses are shown to be either odd, by virtue of having large, extended, or abnormally fecund families, or operating illegally, by building secondary suites. The definition of the norm, of appropriate behavior for the neighborhood, is an important site of contestation. Outlining the normal and the abnormal *in spatial terms* establishes both physical and cultural borders; these bordered spaces can thus be defended on both natural and moral grounds.

Part of the angst of the housing controversy in west-side neighborhoods stemmed from the lack of shared meanings concerning

the natural and the moral. Definitions of normality require inculcation through time and space, as well as the material resources to
back them up. In the case of the Hong Kong Chinese homeowners
who joined the neighborhood, neither time nor superior financial
power could be brought to bear to discipline the new interlopers.
The anger and confusion of many white homeowners manifested
the realization that two powerful visions of the world were in col-
lision, and that the carefully gerrymandered and internalized apparatus of domestic discipline in defining the normal and defending
it from the abnormal was thus rendered ineffective.

Dismay over the inefficacy of old definitions was accompanied
by a growing fear of the economic and cultural power of the new
group. Anxiety over their extended families and the use of "extra"
space in housing reflected a fear of takeover; the newcomers were
quite literally "taking up too much space."[36] The fear of competition over limited resources, such as space in the desirable
Shaughnessy Heights enclave, also surfaced in concerns about
schooling. Many of my Vancouver interviewees felt that the Chinese students were "too competitive," or "too one-track minded."
They worried that their own children, who were more "well balanced," would lose out to this new group scholastically. Several
also expressed dismay at the high cost of expensive English as a
Second Language courses, which they felt siphoned off resources
from after-school activities and art programs.

The fear and anxiety expressed over the loss of neighborhood
character and community also reflected a concern about loss of
control. The meanings of community and neighborhood, once
implicit and secure, were suddenly exposed and contestable. Borders that defined and distinguished the local from the global, in
from out, taste from lack of taste, meaning from meaning, class
from class, race from race, and us from them were no longer stable and impermeable. And the process responsible for these bor-
derline transgressions was inherently difficult for white west-
side residents to criticize. For those who derived their authority
and had long benefited from the dynamism and innovation of
capitalist development, the sudden personal apprehension of its
destructive side was confusing and painful. For example, the
neighborhood of Kerrisdale, which one resident spoke of as
"rather fossilized" when he was a child, was almost completely

transformed in less than a decade. As the buildings fell, so did the presumption of an inherent stability of meaning.

Until the 1980s, large-scale foreign investment in Vancouver real estate was directed primarily into commercial downtown properties. The investors themselves were largely invisible, and urban restructuring was blamed primarily on pro-growth politicians and the collusion between political and development interests. The acquisition of *west-side residential property* by Hong Kong Chinese buyers in the 1980s, however, for the first time opened up the sacred repository of cultural capital and symbolic stability, the home, as a site of investment. While these new spaces of consumption were busily administered to by an army of real-estate "enablers," the Chinese were targeted for marketing and for blame.[37] Negative repercussions of capitalist creative destruction in the neighborhoods were linked with the pending loss of cultural meanings brought about by the entry of this transnational group.

## TRADITION AND THE FOREIGNER

The foreigner is a "symptom"; psychologically he signifies the difficulty we have of living as an *other* and with others; politically he underscores the limits of nation-states and of the national political conscience that characterizes them and that we have all deeply interiorized to the point of considering it normal that there are foreigners, that is, people who do not have the same rights as we do.
        —Julia Kristeva, *Strangers to Ourselves*[38]

To understand how a concept of heritage, or invented tradition, was established and maintained in Vancouver neighborhoods, it is useful to look at the house and landscape traditions of west-side neighborhoods before the conflict. In most, detached, single-family houses predominated. Vancouverites of these early years saw themselves as part of a unique and distinctive urban setting with a high standard of living predicated largely on its mature trees, open green spaces, and single detached houses.[39] They perceived these relatively spacious "models of architectural style" as important alternatives to the tenements and flats common in European and East Coast cities at the turn of the century. In the wealthier suburbs, such as Shaughnessy Heights, their homes could even serve as a bucolic escape, a "foil for work in the industrial city."[40]

In 1986, more than 60 percent of Kerrisdale's houses were detached, single-family homes. (Of the multiple dwelling units in Kerrisdale, 25 percent were located in walk-up apartment buildings of fewer than five stories.) Shaughnessy Heights, in 1979, was more than 83 percent single detached homes, with an extremely low housing density, approximately 2.5 occupied dwellings per acre.[41] West-side neighborhoods were characterized by open, landscaped front yards with mature trees and hedges, and by wide, grassy, landscaped boulevards (see Illustration 9). Typical house styles included the Tudor, the English cottage, and the California bungalow (see Illustration 10).[42]

The ambience of most west-side neighborhoods reflects the English landscape tradition. The gently contoured lawns, scattered clumps of bushes and trees, irregular curved shapes, and natural-looking borders all draw from the landscape tradition of Capability Brown and Humphry Repton, British garden and park designers of the eighteenth century.

ILLUSTRATION 9.  A typical west-side boulevard.

ILLUSTRATION 10.    Vancouver house in British Tudor–revival style.

The use of British architectural and landscape symbols in Vancouver's west-side neighborhoods was deliberate. The Canadian
Pacific Railroad intentionally modeled the elite subdivision of
Shaughnessy Heights on the pastoral myths of a romanticized
English countryside.[43] By claiming an ongoing, albeit transformed,
English tradition in the new spaces of Vancouver, an easy, unhur-
ried grandeur and an idyllic, preindustrial way of life gradually
became recognized and incorporated as the symbols and codes of
the new suburban elite of "British" Columbia. The appropriation
and reworking of these symbols over time operated on an aesthetic level that enabled residents of these neighborhoods to feel
pride of place, security and well-being, and, like the Shaughnessy
graffitist, to assume a universal notion of appropriateness, of what
"fit" the neighborhood. Much of the anxiety of loss expressed in
the late 1980s related not to the actual buildings or trees that were
demolished, but to a more general fear of deprivation and dispossession of this way of life.

An examination of the angry letters to City Hall from 1987
through 1991 uncovers a litany of ambience words—"quality,"

"character," "heritage," "tradition," "neighborliness," and "community." Many letter writers decried the loss of the "village atmosphere" or the "coziness" of the past; nearly all were angry at the destruction of the community's trees and gardens. In a public meeting in 1985, a spokesperson for the Shaughnessy Heights Property Owners Association (SHPOA) expressed anxiety about the burgeoning monster houses (then called Vancouver Specials and associated more with east-side neighborhoods) and urged rapid action on a first set of proposed regulations:

> "You get three or four of these in a couple of blocks," he explained, "and all of a sudden you have completely changed the character of a neighborhood." Executive Secretary of the organization, Evelyn Mackay, said the "threat" of Vancouver Specials had shaken the confidence of Shaughnessy residents in their ability to preserve the character of their area. "This is a real threat to a neighborhood—to have these oversized houses suddenly spring up on a street which we thought was safe and secure forever," she remarked.[44]

These strongly felt sentiments, expressed in terms of heritage, character, and safety, clearly involved more than the construction of monster houses, more than differences of taste or ideas about what was appropriate or ugly. Underlying these expressions of emotional loss was fear for their ability to reproduce a particular way of life. All the perceived negative changes to the values and dispositions of the domestic seemed to be the work of foreigners, who threatened established patterns in multiple ways. Jane Samuels, an upper-class homeowner in Shaughnessy, expressed her confusion and resistance to these outside, or nonlocal, forces in our 1991 interview:

> It sounds reactionary, but I don't think I'm a reactionary person. And I realize that and it's very frightening when you have to sort of start holding on to things. But when you see everything being destroyed and pretty soon it will be sort of treeless, you have the feeling. . . . We live outside in this area, in this city. We don't live inside; it's not an indoor city. We're outside a lot. And people who are used to living in an apartment in New York or Hong Kong don't realize that you can have more green space. It's an appreciation factor that you find out once you've lived here. It's sort of bigger than the way I'm talking. I'm talking in a very small way. I guess there is that large threat that the whole city is changing. . . . Maybe we don't want to make it that attractive [to outsiders], to make it that big. As you watch all these cities grow, you start

saying, *Why* do we want to get bigger? Why? What good does it do all
the people who are living here now?

The nature imagery is not accidental here, but part of a much
broader ideological move that equates domesticity with nature,
and alien bodies or practices with the unnatural, abnormal, and
unhomely, Comaroff and Comaroff suggest: "The ideological
struggle to naturalize the doctrine of domesticity was, from the
first, part of the middle-class endeavor to secure its cultural hegemony. . . . We use the term 'naturalize' advisedly here. The effort
to disseminate the idea of domesticity was saturated with natural imagery." Naturalizing the domestic modern in this way is an
effort to preserve a particular way of life. It is a thread in the
weaving of a "total moral order," where the home and all its associated constructs serve "as internalized mechanisms of discipline
and control."[45]

The eighteenth-century British landscape tradition, the foundation for nineteenth- and twentieth-century revivals, is most
notable for the artificial creation of a "natural" look. In the
designs of Capability Brown and others, landscapes were subtly
but painstakingly altered to conform to an ideal of a perfectly natural, idyllic, and seemingly untouched state.[46] The faith in natural harmony and the logic of nature corresponded with essays by
John Locke that emphasized harmonious and peaceful human
relations as a natural state of social being.

The creation of a natural landscape and the promotion of naturally peaceful human relations in liberal political philosophy
occurred when Britain was dominated socially, politically, and
economically by a small group of great landowners. For centuries
the ownership of land was the surest basis of gaining and maintaining power in England. The power was invested not in the use
of land for farming but in the social relations engendered through
its ownership. The attraction and retention of tenants who would
provide military and political support, allegiance, and rent money
was crucial for manifesting the type of wealthy and well-endowed
establishment that could then attract money from other sources.
In this process, the image of the house was key.[47]

The natural landscape and the social harmony expressed by
landscapers and philosophers of the eighteenth century served to

naturalize the inequitable social relations of landowner and tenant that were part of British preindustrial and early industrial society.[48] They also served as a blind for the true manner in which wealth was acquired and legitimacy maintained. The houses themselves were integral to the continuation of these power relations and the ongoing maintenance of domination by a ruling elite. British housing and landscape styles appropriated for Vancouver's west-side neighborhoods were part of sedimented networks of power that both shaped and were shaped by these historical spatial relations. Although the symbols and meanings of house and garden were obviously transformed in their Canadian regeneration, they nevertheless embodied and retained some of these latent meanings and traditions.

In west-side Vancouver neighborhoods, much of the aesthetic dislike of the monster house was associated with the blurring of the distinctions between old and new, and between old money and new money. The symbols of the nineteenth-century British landscape were based on an idealized lifestyle of the eighteenth-century landed aristocracy. Part of the romanticized image of this pre-industrial setting was also predicated on a disparagement of capitalism and industrialism and the new money and money makers involved in these processes.[49] Although this antibourgeois sentiment faded somewhat in Vancouver, the dislike of new money and its threat to the established meanings of an older elite remains strong. The monster houses, which do not blend in and are brashly new and ostentatious in size and ornamentation, are identified with a nouveau riche display that is antithetical to the retiring and faded-gentry image of old-money Vancouverites.

Statistics from a 1987 market research report showed that on average, 44.1 percent of Kerrisdale residents' total assets were tied up in their houses.[50] This differs sharply from the widely held image and representation of the new Chinese buyers of Kerrisdale houses, who were reputed to arrive from Hong Kong with suitcases full of cash with which they purchased houses outright. I was told an anecdote on three occasions about a Hong Kong Chinese buyer who flew to Vancouver from Hong Kong, was driven through west-side neighborhoods in a limousine, and immediately paid cash for several houses.

Although the anecdote was told with different intentions by different people, it reflected the public representation of the "typical" cash-rich Hong Kong Chinese buyer. And this image was inextricably linked to the recent house demolitions, the construction of the new monster houses, and the loss of neighborhood character. Greed and profit were seen as the evil motives of the nouveau riche, who were represented as caring little or nothing for Vancouver's vaunted traditions and heritage. The conflation of this new wealthy group with the Hong Kong Chinese immigrants was made explicit in several ways. A March 5, 1988, letter to Vancouver's Planning Department commented on a proposal to limit house size: "My greatest concern, which is only briefly mentioned in the proposal, is that of 'neighborhood character.'. . . This is a charming section of Vancouver with beautiful tree lined streets and Georgian, Tudor or Colonial type architecture. However, this is changing with the influx of new people wanting to build these huge eyesores. . . . Let's act now before all of Vancouver becomes another Richmond."[51]

Richmond was a rapidly growing suburb of Vancouver with a particularly high proportion of Chinese Canadian residents, many of whom immigrated from Hong Kong in the 1980s.[52] The letter writer attempted to separate the building activity in Richmond, which was explicitly linked to an "influx of new people," from his own "charming" section of Vancouver. The preference for a particular type of cultural consumption and lifestyle, such as owning a Georgian-, Tudor-, or Colonial-style house is expressed as a manifestation of a taste unquestionably superior to other tastes, such as owning a monster house. Here the expression of difference serves to legitimate and confirm social and cultural differences between the long-term white residents and the nouveau riche Chinese interlopers, and at the same time to legitimate and confirm the systems or structures implicated in the reproduction of these social differences.[53]

The attempt to separate and distinguish old-money values, traditions, and neighborhoods, from the culture and traditions of the new-money immigrants was made repeatedly in opinion pieces and letters to the editor in the local media. Often the traditional values that were being lost were equated with national identity, of what it meant to *be* Canadian, as in a letter to the editor of the local *Western News,* July 26, 1989: "Canadians see monster housing as an arrogant visible demonstration of the destruction of Canadian culture. Yes, we have a Canadian identity and Canadians should beware of persons who say we don't while they try to rebuild Canada in a different mould for their own purpose and profit."

The reference to profit in the letter was directed at the transnational Hong Kong Chinese, widely represented as responsible for house-price escalation as a result of using homes for speculation rather than as places to live. Investing in tasteful or "high" culture in Vancouver society, which includes the home where one lives, secures profit yet *does not have to be pursued as profit.* Living in an established area such as Shaughnessy Heights, for example, purportedly because the character of the neighborhood *feels* right, allows homeowners to profess innocence of any cynical or mercenary motives such as profit, yet establishes their fundamental connection to the underlying systems that generate it.[54]

Although the letter referred to the destruction of a national identity, there was also a clear concern about the loss of individual and group social identity. Profit-generating development in Vancouver's east-side neighborhoods and outlying areas was rarely contested, nor were those areas (which are far more economically and racially mixed) defended on the grounds of preserving heritage, tradition, or identity. The violence of the reaction against the aesthetic changes in the west-side houses betrayed the fear that the symbols of the established and dominant group were being eroded. With that erosion went the chance of appropriating, and *naturalizing* the appropriation of, the rare rights and assets that are dependent on one's position in social space, as well as the distribution of those assets in geographical space.[55] When other dominant class fragments, expressing differing patterns of cultural consumption and lifestyle, gain access to the formerly self-contained and bordered spaces of hegemony, they threaten the distinction of the original group by depreciating old meanings and by calling into question the legitimated "natural" association of real differences in economic relations with differences in taste.

## Transnational Imaginings

I feel that I'm so mixed and assimilated that I cannot distinguish one from another. . . . It's kind of instinctual that when I meet up with somebody who's Western in their values, it doesn't strike me as being strange or unusual, . . . and I think it's because I've just been living and brought up in a Western world for so long that nothing strikes me as being unusual. Likewise, I think, with Chinese cultures. Because I'm sort of just in it. You know, absorbed.

      —Hong Kong Chinese woman[56]

The world culture is created through the increasing interconnectedness of varied local cultures, as well as through the development of cultures without a clear anchorage in any one territory. . . . But to this global inter-connected diversity people can relate in different ways. For one thing, there are cosmopolitans, and there are locals.

      —Ulf Hannerz, "Cosmopolitans and Locals"[57]

By the time I interviewed people in Hong Kong in 1991, the controversy over the monster houses was well known. As one future Vancouver immigrant responded, "You'd have to be deaf and blind not to have heard about that by now." In 1989, Hong Kong television crews flew to Vancouver to film a story on monster houses and tree removal, and to investigate reports of an increasing number of racist incidents involving Chinese Canadians in the city. The television show that was aired in Hong Kong depicted Vancouver as a city struggling with profound change. The show explicitly linked racism, particularly against the Chinese, with the urban battles over changes in the built environment.[58] The news media in Hong Kong also focused on reports of racism, and the English- and Chinese-language newspapers and journals did stories on the monster houses and on the growing anti-Asian backlash in Vancouver in the late 1980s and early 1990s.[59] Finally, word-of-mouth stories about the large houses and the growing anger and reprisals in Vancouver spread the information even more rapidly and completely. By the time I asked questions about house preferences and preferred neighborhood locations, individuals in Hong Kong who were intending to emigrate to Vancouver were well aware of the implications of their answers.

The response in Hong Kong in 1991 to the monster-house controversy was mixed, but most potential emigrants to Vancouver sought to distance themselves from the people who had emigrated earlier and were directly involved in the well-publicized frictions around the issue. Hong Kong interviewees told me that cultural differences—for example, methods of manifesting style and wealth—were probably responsible for many of the incidents that they had seen on television or heard about from friends. In an interview in the coffee shop of Hong Kong's Furama Hotel in April 1991, Joanne Siu discussed the Vancouver controversy with regret. She felt that cultural and economic differences were largely to blame.

> Many of us in Vancouver, we go back and forth now; many of my friends are now going to settle down in Vancouver, they bought properties. We feel that much of the fault is ours. Because sometimes we're inconsiderate. We make a lot of noise; after people have gone to sleep we have a mahjong party. . . . We have upset their lifestyle. We go over and we bring all this money over and it's changed things. Their prices have gone up and it's made the local people find it hard to buy a house, they find it hard to run a business. Lots of things have changed and we have no right. In a way, I expect them to get angry.

Joanne is a wealthy, extremely polished woman who emigrated to Canada in 1967 and then returned to Hong Kong because of an illness in the family. She speaks several languages fluently and travels around the world with some regularity. Her mother lives in Vancouver, and Joanne has visited the city frequently since leaving Canada in the late 1970s. She invested in a "number of properties" in Oakridge during 1982, but planned to sell them and purchase a home where she and her husband would eventually live when they moved to Vancouver in late 1996. In 1987 she found the perfect home and sold the other houses, one for double the price she had paid for it in 1982. Joanne felt that the current Vancouver immigrants from Hong Kong were different from those who had moved to Canada in the 1960s. According to her, the earlier immigrants such as herself were "younger people [who] were going there to settle down, . . . [who] intended to mix in."

As I spoke with Joanne it was clear that she was distancing herself from contemporary Hong Kong immigrants in Vancouver and also offering a rationale for the increasing evidence of racism in the city, which she herself had experienced. After describing one extremely unpleasant incident in a Vancouver shop, where her

use of Shanghaiese with her friend led to their ostracization by
the shopkeeper, she explained that understanding the cultural
motivations (such as anger over the monster house) for such racist
behavior allowed her to feel better about the situation. By con-
flating racialized conflict with antagonism toward cultural difference, she was able to position the friction as something particular and fixed in time and place. She could then remain outside
and above the friction, owing to her particular understandings and
situation as a "westernized" Canadian citizen and cosmopolitan
woman of the world.

   According to Hannerz, the cosmopolitan traveler in world culture is able to manage meaning in ever-shifting and diverse circumstances, and appreciates and appropriates cultural difference
for both survival and strategic purposes.[60] Knowledge of cultural
difference is crucial if the world traveler is to retain some control and even mastery over new nonlocal situations. Racism, as
a process of construction in which difference is naturalized and
fixed, undermines the advantages of cosmopolitan knowledge by
permanently positioning the Other as an outsider—as someone
who, by virtue of skin color, can never fit in. As a world traveler,
a Hong Kong woman educated in English schools who has lived
in Canada, holds Canadian citizenship, and understands Canadian
culture, Joanne's "decontextualized cultural capital" was greatly
threatened by this form of racism. By equating the racism manifested in the monster-house controversy with a fear of cultural
difference rather than the ongoing construction of an outsider,
Joanne was able to position herself, in her own mind, outside the
fray. Her ability to link citizenship with *degree of assimilation*
rather than with race enabled her to "be Canadian" in her own
understanding. As a cosmopolitan woman, she had assimilated,
become westernized, to the point where she felt removed from
the monster-house conflict and the types of racism that circumscribed it. She rationalized the racism she encountered in a Vancouver shop as the residue from a cultural disturbance that would
heal with time.

   Joanne Siu's views of cultural citizenship ally closely with Raymond William's description of national identity as a social and
cultural relationship or pact shaped over time. For Williams, the
conflicts around race and nation in Britain stemmed from the

cultural alienness of immigrants and their inability to share in the "lived and formed identities" of the English working man. If foreigners could only assimilate *enough*, they would eventually become British and be accepted in mainstream white British society.[61] What Williams's argument elided was the element of power and material gain implicated in the production of concepts and categories like race and nation, as well as in the spatial sediment that accrues over time. Gilroy, for example, demonstrated the ways in which race and culture were deliberately conflated to legitimate the exclusion of blacks from "being British" in Thatcher's England.[62] During the reign of the Thatcher government, which relied largely on vulgar nationalist conceptions for political legitimacy, the linking of race and culture and the naturalized sense that white culture was equivalent to British culture positioned black culture and black people permanently outside the possibility of national belonging. This strategy aided the Conservative Party to maintain power during an extended and painful period of economic duress for the poor and middle classes.

For Chinese immigrants in North America, the conflation of racism with antagonism to cultural difference remains strong and has produced a vast discourse of potential "belongingness" based on degree of assimilation. There is always a stereotyped niche available for any group, from the "model minority" or "bridge builder," to the "coolie laborer" of earlier days. The possibility of actually advancing beyond these static, culturally inscribed positions, however, is small. Both Ong and Lowe examined the tenacity of these niches in the United States, arguing that despite the "whitening" effects of money and a culture of assimilation for those at the top, Asian immigrants continued to be excluded from the national political community.[63] Racial divisions and conflicts remain substantial and primary, and the dream of complete immigrant assimilation and incorporation an inaccessible fantasy.

Joanne Siu's personal perceptions of the conflation of race and culture enabled her to imagine an eventual end to the racial conflicts over the monster houses, and also allowed her to speculate freely in the west-side housing market in the 1980s without recognizing her actions as part of the disturbances she decried. As an acculturated Canadian, she did not view her speculative buying and selling of several properties on Fortieth Street as participating in or exacerbating the cultural fracas. Similarly, she saw her proposed construction of a large, three-story house in a west-side neighborhood as a completely separate issue, because her taste was westernized. She said in response to a question about cultural conceptions of space:

> J:  Chinese people in general (when I say Chinese I'm including the Taiwanese) like everything big, grand, show-offish. They want to show off. When I first thought of rebuilding our property, I thought of a two-story house. Then I thought, Well, if I've got this space, why shouldn't I give it the maximum? Part of the reason is because everybody here lives in cramped spaces. Even the well off. If we want to live in a three-story we can't, partly because we know with 1997 it's not worthwhile spending that money. So when we go somewhere where we find that it's going to be permanent, we would like a big house, a garden. It's not practical, but somehow this is just the idea we have, that we want a big place. It's something you always wanted. . . . But I won't build what they call a monster house.
>
> K:  Because you heard about it or. . . .
>
> J:  No, my sense of taste. I don't like that kind of a square thing. . . . I've always been very westernized.

Joanne's reference to "showing off" was reiterated by many people I spoke with in Hong Kong and Vancouver, who felt that the manifestation of wealth in the choice of car or house or clothing was fundamentally different for white Vancouver residents and more recent Hong Kong immigrants. Several believed that this difference had led to jealousy and exacerbated expressions of racism against Chinese people. The manifestation of prosperity, particularly the visible display of name-brand and high-status items, was generally considered appropriate in Hong Kong elite society, where differentiation between dominant class fractions is often predicated on an overt display of wealth.[64]

In elite Hong Kong circles, visible displays of wealth are crucial to acceptance into the dominant group in society. And, as in Vancouver and elsewhere, acceptance into the dominant group is crucial for continued access to the resources and networks of those holding the reins of economic power. The historical legacy of cultural consumption in Hong Kong was described to me by Sally Liu, in a 1991 interview in Hong Kong, as originating in

Shanghai, where maintaining face was crucial for retaining one's place in a particular establishment circle.[65] When I asked about the contemporary Hong Kong cultural style, she described it as a mix of Chinese regional cultures that had formed a distinct hybrid. For her, Hong Kong's unique context in time and place was crucial in the formation of a new cultural identity.

> It's very different now because one becomes very half-merged, the Shanghainese and the Cantonese—and we become very much a Hong Kong-nese. We become very practical people. But, because money in Hong Kong is so easy, has always been easy, so we have a wide section of nouveau riche people. And fiscal wealth is very important. We have, all of us, only been here forty, fifty years at the most. A few families will be over a hundred years. And basically, even though you might be very wealthy in China or you might be very poor, you all became quite poor to begin with because of the war, so you all have to start from scratch. Also, because of the opportunities available, you have many nouveau riche. So I think out of the necessity to survive and out of having no confidence, you have to be very physical, very physical with the car you drive, very physical with the jewelry you have, your address, your house, how to talk big, because this is the way you get into a particular type of circle.

Although neighborhood address is important for manifesting wealth and demonstrating belonging within an elite circle, the house itself is less so. Owing to the scarcity of land and its governmental control in Hong Kong, only the extremely wealthy can afford to live in a single-family detached house. The automobile, rather than the house, is thus the more ubiquitous symbol of wealth and prestige for the upper classes. In the move to Vancouver, however, some attitudes about display and the manifestation of wealth were clearly transferred to the house. Bill and Jenny Leung, a young couple planning to emigrate to Vancouver a few months after I spoke with them, described their visit to a west-side neighborhood where Bill's sister had already purchased a house for them. They claimed that they could differentiate between the houses owned by long-term Vancouver residents and those owned by Hong Kong people. Their criteria for judgment were the types of materials used and the level of money spent on the house by the occupants. Bill Leung and I discussed these issues in an interview in Hong Kong in September 1991.

B: In Vancouver it's quite difficult for us to differentiate the rich class and the poor class, since they all wear the same clothes! They are not showing off. In Hong Kong the behavior is different. When you're richer you use clothes, a precious watch, to show off. The local people in Vancouver, ... they are so simple, not living a luxurious life. The one area that's most obvious is the house. When I look at houses, I can differentiate between whether they belong to the local people or the Hong Kong people.

K: How?

B: The house, for example, the door is a golden color, sometimes the ground is all metal and with glass. And all made with marble. I can see they spend a lot of money. ... Also driving the Mercedes.

Other opinions were expressed in my Hong Kong interviews that corroborated these viewpoints and also indicated nuanced differences in perception that reflected generational, professional, and class divides. Although I interviewed only people wealthy enough to consider moving to Vancouver's west-side neighborhoods, there was a heterogeneity of responses to my questions about the monster houses, architectural design, choice of neighborhood, and symbols and meanings of home and garden. In contrast to the often monolithic presentation of the Hong Kong Chinese in media reports and conversations in Vancouver, the positions of the immigrants were structured and differentiated by multiple and diverse dynamics, which were reflected in a variety of responses.

## FENG SHUI AND THE ART OF LIVING

Aversion to different life-styles is perhaps one of the strongest barriers between the classes. ... At stake in every struggle over art there is also the imposition of an art of living, that is, the transmutation of an arbitrary way of living into the legitimate way of life.
    —Pierre Bourdieu, *Distinction*[66]

For people who believe in feng shui, or geomancy, alterations to a landscape by human beings can cause disturbances that "may redound to influence, even control, the fortunes of those who intrude."[67] Careful attention to *qi* (cosmic energy or life breath) is crucial for ensuring that the landscape alteration is a positive one, ultimately beneficial for the health, well-being, and financial outlook of the person inhabiting a particular place. The siting of a new building, including the entrance and method of circulation, affects the flow of qi, and is thus an important factor for feng shui followers to consider. The directions north, south, east, and west have differing meanings and, depending on the local landscape features, can be positive or negative points of entrance for a harmonious balance of yin and yang and an unhindered flow of qi.

Many immigrants whom I interviewed believed in the principles of feng shui to some degree, especially in the general situation of the house and in the layout of certain landscape features. These beliefs reflected a different interpretation and ordering of space from those of most older Vancouver residents, although both groups expressed a desire for remarkably similar attributes relating to the house's situation. Because these nearly parallel desires employed a different conceptual schema, however, they were a source of some discontent, and were used by a number of white residents as examples of the irrational and unnatural spatial imaginings of the Chinese immigrants. House and garden situation and style thus became caught up in larger narratives of the relative "rationality" of the Chinese, posing new barriers to acceptance within the larger social and political community.

Bob Iu is a young professional architect in a small, family real-estate company. He was born in Hong Kong and studied in Canada at the University of Manitoba. He practiced as an architect in Vancouver for two-and-a-half years and recently returned to Hong Kong to join the family business. When he worked on a family development project located in downtown Vancouver, he was confronted with some of the traditional ideas his father held regarding the importance of geomancy. In an interview at his Hong Kong office in April 1991 he spoke about the ramification of these differences for the new project:

> My father went and looked at the feng shui of the place. When I look at the site, I evaluate it according to floor space and relationship to area and facilities, . . . very academic. So we started designing the project. Then my father came in with the feng shui and influenced the positioning of the tower. The entrance to the building—he doesn't like it because he doesn't like. . . . I'll show you [he points to an architectural model]. In the model there it's bounded by two streets and basically the only entrance is from this side. And my father doesn't like it

because it's facing west. And according to feng shui it's not good. It had to face north. So what happened was we put in a stair towards the north so that you can actually come into the building from this side. . . . That's how the older generation influences us.

Although many long-term white residents of Vancouver spoke with scorn of the double doors that became ubiquitous on the newly constructed monster houses in the late 1980s, by far the most controversial feature of feng shui in Kerrisdale and Shaughnessy was tree removal. In feng shui belief, trees are tied into qi in complex ways and can represent either good or bad luck, depending on their placement. If trees are poorly placed, either directly in front of the house entrance or windows (blocking qi), they are considered destructive and oppressive.[68] The removal of large trees in monster-house lots was a source of great anger in west-side neighborhoods and led to the formation of social movements specifically established to curtail tree removal in Granville and Kerrisdale. Joanne Siu discussed the negative meanings that a three-stemmed tree on her property held for some Chinese people:

The Chinese have a lot of superstitious ideas about numbers, especially Hong Kong people, and about feng shui. I have a property in Vancouver. There's a birch which has three stems right outside my window. I don't mind that because I love trees and I think it's a beautiful thing and it must be, what, fifty years old. But most Chinese, with three stems, they would say it is three incense sticks and that would be a very negative thing. So they would take it down. And they would do that in a park, which is not their own. There has been talk about people putting poison into the tree so that the tree would die and they have to cut it down. In Vancouver they've been saying things like that. I don't know how true that is, but if I know the Chinese here, I would believe it. I'm from Shanghai and we are less superstitious.

The fortunes of past residents were also frequently believed to have an effect on the prospects of future inhabitants. For those holding this belief, it was important to know the complete histories of houses and their previous tenants, or, if this was impossible, to purchase a newly constructed home. This perceived preference for new houses was a key factor in both the demolition of older structures by developers and in the racialization process that became coupled with the demolitions. In this, as with the other controversies, each set of actors positioned itself as innocent of racism and pointed to either cultural differences (which again they generally deflected to "others," as Joanne did) or to market imperatives to explain their actions.

For example, Jack Sherman, a Vancouver developer who owned a small company and had been involved in construction for thirty years, described mammoth changes in the development industry and in the type of buyer he served in the 1980s. After the recession in the early 1980s, when some developers lost everything, the market remained steady for a few years. In late 1984 and 1985, however, "things started to move," and he began to invest in housing on Vancouver's west side, he said in our June 1991 interview. "All the action was on the west side, and 99 percent of it was triggered by foreign, mostly Hong Kong and Taiwan investors. . . . For a few years, all of my clients were Chinese."

Jack began to change his development and marketing practices "drastically," starting with a greater emphasis on size, and a commitment to building to the maximum square footage per lot. Other changes that he instituted were an increase in the number of bathrooms, a greater emphasis on the entrance, which was now "usually out of marble," a larger kitchen, and a grander en suite master bedroom. When I asked Jack why he made these changes, he told me he was responding to the demands of Chinese buyers. According to the statistics of previous sales, the larger houses with more bedrooms were selling better. Following the logic of numbers, he thus "built to that demand." Jack reported that other development companies were reacting in the same way, because "builders know each other, stick together, and share information."

Max Ng, a Vancouver developer who grew up in Hong Kong and commutes regularly between the two cities, expressed a similar viewpoint in our November 1991 interview in Hong Kong. He emphasized the role of the developer in determining what the average taste of the client was, and then building to that taste. At the same time, he felt convinced that developers had accurately discerned the "Hong Kong Chinese taste," which could be generalized as one requiring lots of space. In response to a question about whether Hong Kong clients were demanding larger, newer houses, or whether the developers were building and supplying larger houses (and creating demand for them), he said:

"Actually, the developer. Tailor-made for the customer. Depending on the taste of the customer. If you live in Hong Kong, if you grow up in Hong Kong, you know space is very important. You have to spend a lot of money to. . . Space is not for free. In Canada, space is free. So when an Asian moves to Vancouver, they like to have more space. Inside or outside. They would prefer to sacrifice the outside space so that they can have more inside space. So this is the need of the people."

According to these developers, the aesthetic tastes, spatial needs, and cultural style of Hong Kong clients influenced house design and the marketing of real estate in Vancouver's west-side neighborhoods. Both also acknowledged the role of the developer in systematically elaborating on perceived cultural differences, and in reacting to them in a manner calculated to derive maximum profit out of the venture. The cultural differences of some immigrants, such as belief in feng shui or dislike of gardening, were seized upon, fixed in place, and processed by the meaning machines of the development industry. The ongoing process of cultural negotiation and change involved in the transition of meanings and systems of belief from one geographical location to another was thus reified and rarified by those able to use this process for profit.

It was clear from my interviews that within the dominant class fractions of Hong Kong there were greatly differing perceptions of preferred housing location and style, and also of the monster-house controversy itself. In a November 1991 interview in Hong Kong, Cindy Liu, a nurse, expressed great anger at the arrogance of the very wealthy Hong Kong immigrants who "park anywhere, get parking tickets, and don't mind." Her aunt owned a house in Richmond and one in Kerrisdale, but Cindy considered herself and her family to be of average income. When she moved to Vancouver she planned to buy a new large house in Richmond because she felt that houses there were spacious, uncrowded, without large gardens to tend, and less "conspicuous" than houses on the west side.

Janet Yo, a highly successful professional artist and the daughter of a wealthy Shanghai banker, was disdainful of the wealthy Hong Kong immigrants in Vancouver for different reasons. She characterized the Chinese in Canada as "very bourgeois, very bourgeois, very low-class," in our December 1991 interview in Hong Kong. She distanced herself from these immigrants by expressing dislike of the monster-house aesthetic and of the mentality of "these rich brassy people coming in, uprooting everything, and digging up this and that." She said that if she emigrated to Vancouver, she would prefer to live in the "British properties in the hills."

In conversations with me, both women sought to distance themselves from the negative monster-house altercations in Vancouver, which they portrayed as partly the fault of earlier Hong Kong immigrants. Although they were from different class fractions, neither identified with the wealthy Hong Kong immigrants, whom they perceived as disrupting Vancouver society and making their own future integration more conspicuous and more difficult. Janet, who came from an elite old-money Shanghai family, decried the Vancouver residents' wholesale application of a nouveau riche mentality to Chinese immigrants. Like Joanne Siu, her status as a cosmopolitan Chinese woman was threatened by the new associations being made about the Hong Kong Chinese as a group. One way she sought to distinguish herself from this group was to express a preference for older "British" properties rather than the new, ostentatious, and "bourgeois" monster houses. Cindy, a member of the Hong Kong middle class with some wealthy family connections, expressed dislike of the arrogance of the earlier Hong Kong immigrants, which she felt exacerbated existing racism in Vancouver society. Her preference was to locate in Richmond, where she believed there were more Hong Kong middle-class immigrants and she could remain inconspicuous and adapt. She felt that the earlier "flashy" group of immigrants behaved in a manner that was "bad for everybody."

In deciding on a residence, the access to a particular style of life and type of neighbor was considered as important as the house itself. Andrea Wong, a nun and professor at the University of Hong Kong whom I interviewed in May 1991, expressed an interest in living in Richmond, which she felt had more space but was less crowded for the price than other areas. She said that the most important factor in choosing a neighborhood was the quiet and the neighbors. "Buying a house in Vancouver, it must be reasonably quiet. And the neighbors should not be causing trouble. Intellectuals. With a certain standard of education."[69] Betty Li, a forty-three-year-old nurse, also referred to "well-educated" neighbors in our April 1991 interview in Hong Kong. Betty said that she didn't want to live on Vancouver's east side because there were "lots of Vietnamese" there. Similarly, John Sinn, a businessman who had bought a property on the west side, listed one of his reasons for choosing a west-side neighborhood as the absence of black people.

The desire to live in a neighborhood with a specific race or group or class of people, and without others, reflected the association of neighborhood power and status with the status and power of its residents. For Andrea Wong and Betty Li, living amid intellectuals with a high degree of education assured them the neighborhood status and atmosphere they desired. The presence of Vietnamese or black people in a neighborhood represented a negative value that they wanted to avoid. The racial classifications employed here exemplify a long-distance, shorthand ranking system that some future immigrants used to determine the best neighborhoods in which to locate. From their own racist categorizations, as well as from their preconceptions of Canadian society, Betty Li and John Sinn sought to determine the best neighborhoods through the exclusion of what they read as "undesirable" groups of people. Since Vietnamese and black people were largely excluded from west-side neighborhoods, these areas were conceptualized as safer, cleaner, and more prestigious.

Encoded in Vancouver's west-side neighborhoods were meanings of class, power, and prestige that were deciphered from across the Pacific and deliberately borrowed or appropriated by the people moving in. Symbols and meanings of distinction were perceived in racial exclusion and neighborhood address, but were considered less important in house and garden style. With respect to the house itself, size, rather than style, was the major consideration expressed by most of my Hong Kong interviewees. Vancouver meanings that related to address, educational priorities, and the exclusion of undesirables were easily understood and readily appropriated by most Hong Kong immigrants. But British eighteenth-century landscape styles embodied in house and garden held little meaning for them.

The choice to buy in west-side neighborhoods reflected the articulation not only of Vancouver and Hong Kong racism, but also of capitalism. Skyrocketing house prices in Kerrisdale and Shaughnessy made homes in these neighborhoods attractive as investment opportunities, as well as places to live. For people intending to emigrate from Hong Kong before 1997, the dual advantage made the housing in these areas difficult to resist. Sam Yip, a wealthy businessman who lived in Jardines Lookout, Hong Kong, purchased a house in Kerrisdale because he felt that it was an undervalued location that would increase in value. He spoke of the house in terms of its potential for profit, but at the same time, he noted that it was also a convenient and safe location for his daughter to live while attending the University of British Columbia.

Sam's strategy in the purchase of a Vancouver house was complex, yet he spoke to me of the profitability of the venture as his first consideration. Making a large or quick profit on the purchase of a house or flat is seen as a positive business ploy that one can speak of with pride and accomplishment. Speculating in the housing market is a common practice in Hong Kong, where political and economic factors produced a favorable environment for collecting fast profits in the development and real-estate industries. Unlike Vancouver, where housing has resonance as symbolic capital, housing in Hong Kong is generally perceived as a commodity to be bought and sold for a profit, much the same as any other commodity.

The impermanence of the built environment in Hong Kong undoubtedly had an effect on Hong Kong residents' perceptions of house and home. Most of the people I spoke with there felt little emotion about their current residence, but described with enthusiasm the greater amounts of space they would have after moving to Vancouver. Their image of the house itself was clearly secondary to the amount of space it contained, and to the lack of crowding it promised. When Bill and Jenny Leung described their newly purchased house on Forty-seventh Avenue in Vancouver, they emphasized size, separateness, and materials. Bill said to me: "The house in Vancouver is about 10,000 square feet. It's two floors. It's separated from the others. It's about twenty years old and all made in wood. . . . In Hong Kong we live in a flat, not a house, and the area is about 700 square feet."

## Conclusion

I could tell you how many steps make up the streets rising like stairways, and the degree of the arcades' curves, and what kind of zinc scales cover the roofs; but I already know this would be the same as telling you nothing. The city does not consist of this, but of relationships between the measurements of its space and the events of its past.

—Italo Calvino, *Invisible Cities*[70]

A vague but persistent sense of decline in the quality of domestic life in the West has been expressed in books, politics, and popular culture over the last several decades.[71] In Vancouver, British Columbia, this uncomfortable feeling of change and loss was both explicitly and implicitly associated with the arrival of Chinese immigrants from Hong Kong, whose visible purchase and occupation of expensive houses in rapidly transforming neighborhoods positioned them as easy conduits for the anomie and anger incited by the changes. Decline in the quality of domesticity was perceived and linked with this group in a number of ways. Often expressed through metaphor and allegory, the decline of the land, the home, the family, and the nation was associated with the arrival of the outsiders, whose racialized and formerly colonized bodies threatened the established hierarchies of generations.

One of the most pervasive images of invented domestic tradition in the West is that of the cheerily successful, white middle-class nuclear family, with the father at its center as the economic provider, the mother as the locus of social reproduction at home, and two children.[72] Never mind the selective memory necessary to uphold this image and to eradicate those of the servant, the cook, the clerk, the teacher, the nonwhite, the immigrant, and all the others whose depressed wages buttressed this image. This was (and remains) a powerful narrative of domesticity for many Western societies, including Canada.

It is only the uproar occasioned by the naming of David Lam as lieutenant governor, oft-told anecdotes about Chinese entrepreneurs in limousines, and books and media stories about the arrival of the "China tide" that reveal the depth of feeling concerning the possible unseating of this white patriarchal family scene. The surging "flow" of capital allows the inundation of the

fixed landscape and the fixed identities of white middle-class Vancouver residents with new and contaminating meanings. Numerous media articles, popular books, jokes, and anecdotes from the
late 1980s emphasized the threat of invasion or engulfment in this
manner. Words relating to water, such as "tide" or "wave," were
used systematically in reference to the new business activities
and immigration of the Hong Kong Chinese.[73] The metaphor of
destruction and engulfment by water has been shown by authors
like Theweleit, in his study of the German Freicorp, to have been
a symptom in writing and fantasy that related to deep fears about
dissolution and the transgression of boundaries. Theweleit related
these anxieties primarily to concerns about the wholeness and
stability of identity and masculine sexuality.[74] In the Vancouver
case, however, I believe they are also intertwined with fears about
white middle-class economic decline.

The *Vancouver Sun* in 1992 published a series of front-page
articles by reporters Daphne Bramham and Gordon Hamilton entitled "The Death of the Middle Class." The series was prompted
by evidence of a serious decline in wages and standard of living
for middle-income workers in British Columbia in the late 1980s.
The articles chronicled job losses, wage cuts, rising house and
land prices, inflation, and other factors that affected Canadians,
arguing that an estimated 275,000 families (about 800,000 people)
had disappeared from the ranks of the middle class in Canada
*since 1982.* Although a few of these families rose to join the high-
income earners, a far greater percentage experienced a decline in
family income.

> Underlying countless personal tragedies is a restructuring of the Cana-
> dian middle class. The middle class as we knew it is dying; for some
> the dream is already dead. And for some, it has never been better. The
> redistribution of wealth that characterized the post-war years has
> reversed. A polarized economy peopled with many who make less and
> some who make more is emerging. Companies have been restructured.
> Jobs have been restructured. Taxes have been restructured. World trade
> has been restructured. In effect, your life has been restructured.[75]

The polarization of rich and poor increased in Vancouver following the rise of neoliberal policy and concomitant attacks on
social reproduction in the city. In most Western societies of North
America and Europe, the 1980s witnessed a "great U-turn"

unprecedented in the degree to which a relatively prosperous and expanding middle class began to experience the sharp shock of declining wage power, job stability, and the protections afforded by commonly accessible social services and benefits.[76] The newly constituted flexible worker of flexible capitalism felt this decline most immediately and directly when it was quite literally brought *home* into the formerly "protected" spaces of a modern domestic sense of order.[77]

In a key distinction from earlier theoretical works that linked domesticity and modernity in the context of empire, it is not domestic *modernity* which trumps the spaces of premodern home life. Rather, a kind of *hypermodernity* associated with fast capitalism, multiple households, and flexible strategies of home-making is perceived to trump the modern. The new, highly feared subjects of late capitalism are linked with the neoliberal market in an even more intense relationship than the modern subject, and perceived as multiply fragmented in their lifestyles and ways of being in the world. The accelerated economic component of home ownership—manifested in speculation, flipping, and other for-profit housing ventures—was directly associated with the Hong Kong transnationals, who were subsequently positioned as the primary "vectors" of fast capital. At the same time, the declining power of the white middle class in Canada was linked with the same processes of laissez-faire, neoliberal capitalism perceived to have facilitated and perhaps even created this "Asian" home buyer.

In her study of domestic individualism in America, Brown cited the introduction of the market and the rise of domestic ideology as intertwined processes which updated and reshaped individualism.[78] The domestic emerged alongside market-economy expansion and helped shape male and female subjectivity by providing them separate spaces and spatial practices. Thus spatial transformations associated with the home and an increasingly gendered narrative of domesticity shaped and were shaped by the increasing connections with and subjection to the market. The home, in this context, facilitated market expansion in the external, industrial economy. Through serving as a refuge from the socioeconomic uncertainties of the market, and through the profound gendering of productive and reproductive labor, the home was a key site through which individuals were enlisted into modern capitalist modes of subjectivity.

This home world, however, was predicated on specific kinds of spaces and associated practices. To serve as both a "refuge" and a "female" space, the home itself had to be stable and secure. It had to be the Other to the public, male, and frenetic nature of the marketplace. When the market engulfed the home itself, these gendered associations were greatly disrupted, alongside the economic ones.

Few scholars of the domestic have looked at the impact of accelerated spaces, or new kinds of time-space compression, on contemporary understandings of the home. The gendered assumptions of old-style Western patriarchy were established through a market and a mode of modernity with its own highly specific spatiotemporal rhythms. These rhythms have greatly altered with the global economy, leaving normative codes of home and home life unmoored. As the middle-class home itself was drawn into the accelerated speculative pace of laissez-faire global capitalism, established values were thrown out of joint. This disruption became even more evident and upsetting for long-term wealthy white residents because of the unfamiliar values and strategies of those buying and selling the homes.

A common strategy for many Hong Kong transnational families in Vancouver, for example, was to situate the family in multiple locales to take advantage of the different types of resources available in diverse locations. Often the male head of household remained to work in Hong Kong, while the wife and children settled in the new house to establish Canadian residency or attend school. The husband/father would travel frequently back and forth between Hong Kong and Vancouver, and occasionally the wife would return to Hong Kong with him for several months, leaving the children either alone in Vancouver, or cared for by relatives or friends. This multiple siting of the family across space is a new strategy emblematic of the dictates of late capitalism for both capitalists and many migrant laborers worldwide. As Ong has discussed: "'Flexible citizenship' . . . denotes the localizing strategies of subjects who, through a variety of familial and economic practices, seek to evade, deflect, and take advantage of political and economic conditions in different parts of the world. Thus, we cannot analytically delink the operations of family regimes from the regulations of the state and of capital."[79]

The tactical practices of the "astronaut" families located in/between Hong Kong and Vancouver "make sense" as elements of a broader strategy of incorporation into and evasion of various capitalist and nation-state regimes of power. They are, furthermore, practices which have had tremendous ramifications for gender relations and patterns of kinship and home life for the transnational subjects themselves, as Ong and others have documented.[80] What I endeavor to show here, however, are the ways in which these transnational spatial practices disrupted the normative associations of the liberal domestic modern in Canada, in both economic and racial terms.

The intertwining of racial and economic relations reflects the historical production and maintenance of the Canadian domestic, which signifies *modernity* with respect to the imperial project, *refuge* with respect to market uncertainties, and *heimat* or "homeliness" with respect to being in the world—a profoundly patriarchal, racist, and nationalist spatial imagining. The ordering and reordering of space as an ongoing hegemonic formation was threatened by the production of and movement through space of the transnational Chinese home buyers, who were depicted as dangerously "ungrounded" in their national allegiances and "uncanny" in their use of domestic spaces.[81] They resisted the spatial orderings of the home and, by extension, the nation through the movements of their highly codified bodies into formerly discrete spaces, and through their alternative and unfamiliar spatial practices and imaginings.

The home, as the locus of power and control in many colonial narratives, represented the modernizing achievements of colonialism and the transformation of "wild" frontier territory into the domestically managed modern metropole. The loss of white control over these homely spaces rendered unviable many of the established but unspoken codes of both liberalism and modernity—especially those concerned with the separation of spaces into public and private, male and female, white and non-white, colonial and colonized, modern and traditional, and organized and chaotic. The national narratives of liberalism and modernity, as reflected through the spatial organization of house and home, were thus shown to be both unstable and highly contingent.

The free circulation of capital, including real-estate capital, is one highly visible practice associated with the neoliberal changes of the contemporary Canadian state. Neoliberalism, as a philosophy and practice of unfettered global movement, exists in tension with national narratives of permanence and stability and the spatially sedimented histories that associate domesticity, whiteness, and empire. Thus, as the discourse of domesticity is disrupted by new narratives of home and new values and dispositions associated with home life, there are increasing moments of tension—not just between immigrants and long-term residents, but between neoliberal state practices of privatization, speculation and entrepreneurialism, and socially liberal national narratives of coherence, stability, and modernity.

# 6 Conclusion

## The Urban Spatial Politics of Liberal Formations

LIBERALISM AND NEOLIBERALISM take multiple forms, and are understood and experienced differently depending on the context in which they play out. Actually existing liberalism as it is now experienced and activated is based in a sense of a social liberalism deriving from Lockean principles of individual freedoms *and* Keynesian assumptions about state interventions and responsibilities. It serves as a useful narrative framework with which to back up claims to space, to citizenship, to reasonably priced prescription drugs, to multicultural classrooms, to migrant rights in Canada and elsewhere. For the residents of many modern Western nations, a socially liberal cocktail of individual rights, economic redistribution, and state benefits is far more "real" than the abstracted "pure" forms of liberalism or neoliberalism pursued in much of scholarly debate.

Recognizing the context in which abstract ideas are shaped and developed is crucial in a number of ways. It necessitates a more honest engagement with the material spaces and stories of history, the micropolitics of an existing world in which individuals and groups actively manipulate their surroundings. As a result of this engagement, philosophies such as liberalism can be recognized as more than abstract political treatises on social organization and state-society relations, but also as less than blunt technologies of government or instruments of ideological control wielded by state actors. *Liberalism is not a thing, but a formation*, produced and reproduced by multiple actors contingent on the social, political, and economic processes active in society at a given moment. Recognizing the importance of context also increases intellectual openness to the possibility of variations on a theme, as different historical and geographical milieus spawn vastly disparate reworkings of liberal narratives of person and of nation.

The great gaps between a strict Lockean economic liberalism, a social and nationalistic liberalism of midcentury scholars such as Dewey and Marshall, a redistributionist liberalism of Rawls and his followers, and the multicultural liberalism of contemporary figures such as Taylor and Kymlicka indicate just a few of the variations that have played out in different settings. Of course intellectual debate among the followers of different liberal traditions is not a new phenomenon. What I have tried to show here, instead, is the deep overlap of liberal ideologies in the everyday social world. This world, moreover, exists in real time and space; in it, the quotidien struggles over resources and way of life affect, and have always affected, the formation of liberalism.

The crucial importance of actual *space* in political identification—of territorial allegiance and its impact on the formation of political communities and ways of thought—is conspicuously absent in the scholarly writings of both historical and contemporary varieties of British liberalism. Indeed, as Mehta points out, "political theorists in the Anglo-American liberal tradition have, for the most part, not only ignored the links between political identity and territory, but have also conceptualized the former in terms that at least implicitly deny any significance to the latter and to the links between the two."[1] Mehta's reading of the critical absence of space in liberalism, however, reproduces the mistake of most political theorists in reading spatial organization as autonomous from the world of ideas. In this view, Locke's *initial* theoretical separation of a notion of territorial identification from the political world *then* allowed Locke and other liberal thinkers to endorse the British Empire. Because of their *lack of a conception* "of belonging or territorial togetherness, liberals were unable to recognize and appreciate the political integrity of various nonconsensual societies."[2] But a more dialectical reading of this case— one which brings lived space in as mutually constitutive of social organization rather than reactive to it—would locate Locke's theories *themselves* as at least partially reflective of his contemporary material world. One could argue that it was not through the basic lack of a conception of "territorial togetherness" or an inability to appreciate the "political integrity" of colonized societies which then *allowed* the early liberals to accept and even advocate a nonconsensual and nonuniversalist colonial rule; rather, it was a deliberate elision of these territorial equations. Inclusion of them would have disrupted the hierarchical stratification premised on British cultural and national distinctions of which Locke and his upper-class white male followers were substantial beneficiaries. He did not write in a vacuum, nor did his theoretical disciples; the world of British Empire and all its associated narratives of modernity and racial science affected their thinking and their quality of life, and was reflected in their writings. This type of conceptualization rescues both space *and* liberalism from the usual passive roles they are given in political theory, and creates a more dynamic and dialectic vision of liberal formations.

In the transnational migration and urban change discussed in this book, "territorial" space is a key marker of political belonging used by both the immigrants and the long-term residents of the city. The latter invoked images and stories of spatial identification with particular landscapes and houses, and yoked these both explicitly and implicitly to the right to membership in the political community; the newcomers actively contested these narratives. They did so by exposing the British-inflected and racist cultural norms of historical Vancouver spaces, and also by practicing (and occasionally advocating) alternative dispositions of home, landscape, and territory. It is in the ongoing encounter and dispute between these two groups that liberalism is actively reworked, and it is in the context of the city and its territorial spaces and allegiances that this reformulation occurs.

Thus a main premise of this book is that the urban milieu is both a fascinating and an important arena in which to examine the spatial politics of liberal formation in the contemporary era. The struggle between a socially liberal conceptualization of state-society relations, and the neoliberal variation that presents its greatest current challenge, is evident in the strategies of positioning played out at all levels of city politics. The demand for the necessities of social reproduction, striated by class, is also a demand for the rights to urban consumption. These demands have become strident and insistent in recent decades, as cities have been starved of funds by neoliberal policies at provincial and federal scales, and as class polarization has bled the poor, working, and even middle classes of the necessities of social reproduction. The attendant sense of anxiety and loss, while experienced most desperately by the poor, circulates in unusual ways through the ranks of some dominant class fractions as well. For those who have relied on the cultural distinctions of place for status and for access to certain kinds of resources, the neoliberal swing of the last few years has meant a declining ability to retain spatial privilege and to buttress that privilege with the "normal" liberal narratives of the rational. Thus liberalism has become an agitated and mutable formation in recent years, and I believe that it is in the struggle over the rights to the city, in the broadest sense, that the future shape of liberalism and neoliberalism is now being wrought.[3]

Transnationalism is a key piece of this puzzle in many urban milieus, because of its inherent subversion of established patterns of social and spatial organization. Most transnational migrants land in cities and, through their actions and their reception, shake up conventional categories and the narratives and spaces which uphold them. Transnationalism, in this sense, dislodges older assumptions and introduces new ones. In so doing, it disturbs normative patterns of life encrusted in tradition and memory that aided in the ongoing hegemony of dominant class and race interests.

The movement of transnational migrants between Hong Kong and Vancouver, for example, challenged the hegemonies of race and class in multiple ways. But this challenge was vastly complicated by the ways in which these hegemonies were bound up with the ongoing formation of, and constant struggle over, liberal definitions and meanings. In many modern Western nations such as Canada, hegemonic formations about race and class were formed in the crucible of British liberal ideas about the individual, the modern, and the rational. For liberal thinkers, rational individuals must be guaranteed as a universal right the opportunity to participate in political life. Without this claim to universalism, the society cannot be considered an enlightened modern nation. Implicit in the concept of the rational, however, are cultural markers that denote acceptable forms of rationality—markers crucial in either guaranteeing or denying acceptance into the universal membership of the political community.[4] Levels of acceptance, moreover, were and remain multiscaled; they range from claims to national citizenship to those of full inclusion in neighborhood zoning discussions. And acceptance is also multilayered, ranging from formal indicators of acceptance and inclusion (e.g., to national citizenship) to bands of distinction that are implicit and unnamed, the "informal" layers of (un)acceptability.[5]

Race is often the great unnamed that floats between these layers, disturbing the universal claims of Enlightenment apologists and unsettling the easy self-righteousness of social liberals. Immigration, which usually involves the encounter of strangers considered racially Other, forces the issue of race to be named and broached discursively and materially. This naming thrusts into the limelight the informal, the implicit, the assumed, and the "taken-for-granted" notions that have upheld white hegemony for centuries.

Furthermore, immigrant encounters are crucial in the ongoing constitution of racialized (and classed) groups such as the Chinese migrants in Vancouver, as well as their white counterparts. As Isin has discussed with respect to citizenship formation, it is through the process of encounter and categorization that group identities are made.[6] Thus racialized groups are both formed and continually reconstituted through migrant encounters and citizenship formation, and each point of contact leads to new kinds of destabilization for existing categories of meaning.

For wealthy immigrants—particularly for wealthy transnational immigrants, who are often perceived as belonging to a globalized world of movement and nonallegiance to place—the issue of race is intertwined with the issue of class, and both are linked with the constitution of the liberal nation. The racialization process, which involves immigrants and nonwhite groups such as natives and former slaves, has always been a formative process for nation-building in the West, aiding in the establishment of boundaries and categories, as well as the constitution of groups.[7] This process, however, has become more complicated in the last few decades, as mobile migrants and flexible citizens disrupt the "standard" categories and binary conceptualizations of home and away, inside and outside, and citizen and sojourner. Indeed, the subject of "citizenship" has become a vital topic in both scholarly and popular debate in recent years precisely because of the ways in which new forms of migration and migrant life are challenging familiar notions of the nation and its ongoing constitution.[8] Alternative conceptualizations of both (deterritorialized) national allegiance and national political life, moreover, have become more insistent and disruptive because of the dominant class positions of a new global class of "flexible citizens," such as those discussed in this book.

Contemporary socially liberal philosophies of national belonging and allegiance, such as the set of ideas associated with citizenship and multiculturalism, are particularly affected by transnationalism and neoliberal globalization. These ideologies, among others, defined an image of the nation and of national development premised on a territorial understanding of national identity and allegiance actively controlled and monitored by the technologies of the state. The discourse of social liberalism, as manifested in philosophies such as multiculturalism, rests on both the formation of the nation and the policies of the state. The discourse of neoliberalism, by contrast, depicts a warm, planetary embrace of global humanity, existing and interacting across national borders in a friendly, historically void, and geographically featureless abstract space. This neutral level playing field, moreover, is imagined as autonomous from the state, regulated only through the perfect functioning of the global market. Although this is a patently false image, belying the ongoing state intervention necessary to produce "marketable" spaces of this kind, it is quickly becoming dominant. In this context, citizenship, multiculturalism, and numerous other liberal narratives of national belonging are slowly morphing into new formations.[9] The profound war of positioning between these two forms of liberalism and their associated narratives and spatial foundations is currently being waged in Vancouver and in many other postindustrial cities.

This sense of the complexity of the war and the importance of the battles makes me resist the seductive assuredness of readings such as Bourdieu and Wacquant's on multiculturalism. They condemn the theoretical discussions around "liberal" issues such as multiculturalism as the NewLiberalSpeak "neo-babble" of intellectuals, who ought to stop wasting their time and return to the "important" issues of "capitalism, class, exploitation, domination and inequality."[10] I find their stance both simplistic and politically unhelpful. Like all liberal philosophies, multiculturalism is unquestionably and profoundly linked with capitalism. However, this linkage is neither static nor uniform, and comprehending the nature of its ever-changing formation can tell us something about both multiculturalism *and* capitalism. Further, examining how liberal philosophies such as multiculturalism are produced through the interaction of state actors, capitalist interests, and actually existing people gives insight into hegemonic forces that are broader, more complex, and more accurate than purely economistic readings of the social world.

The subtle and not so subtle differences between liberal philosophies matter—for the constitution of political formations, as well as for theoretical rigor. But these are precisely the kinds of nuances that are swept away by macroanalyses and broad theoretical claims unsullied by the banalities of everyday life. Much of the currently fashionable governmentality literature offers a fascinating and useful portrait of subjectivity formation in the neoliberal era. Yet despite the brilliance of these analyses, I find the general tenor of writings in this genre top-heavy; neoliberal state ideologies seem primarily to *work on* formed subjects, rather than state-subject relationships being dialectically constituted. I've attempted to show here the multiple ways that subjects actively manipulate their world, using whatever narratives to hand that might help them obtain the urban resources they want or need. These subjects are constituted through this struggle, but they also shape the various narrative strands of liberalism and neoliberalism at the same time.

Liberalism's constant "failure" to live up to the promise of universalism is not merely an accident of implementation. It is a key part of the constitution of the liberal nation. As scholars such as Mehta and Pateman have shown for liberalism, and Isin and Lowe for citizenship, the logic of alterity is *immanent to* these conceptual apparati rather than external to them. Citizenship and liberalism historically have been shown to foster not equality and inclusivity but internal differentiation and hierarchy. They do not progressively include, but rather produce otherness as part of the process of defining what counts as political.[11]

There exists, however, a "promise" of social liberalism. The belief in state intervention and redistribution is a key branch of contemporary twentieth-century liberalism, as are understandings of fundamental individual and group human rights. A sense of these "human" rights now crosses international borders and provides the discursive rationale and backbone for universal standards of human dignity and social justice globally. In the face of the growing power of neoliberal discourse, this important promise should not be abandoned wholesale. Perhaps the pretensions of social liberalism and enlightened modernity toward the principles of universality have been irrevocably damaged, "punctured from the moment of their conception in the womb of colonial space."[12] Nevertheless, I am hopeful that some vestige of the promise remains, and contains within it the potential to galvanize political action in forms other than nationalist "camps."

I return, with a warning, to the ardent neoliberal vision of globalization with which I began this book: "Their image of globalization offers the promise of a unified humanity no longer divided by East and West, North and South, Europe and its Others, the rich and the poor. As if they were underwritten by the desire to erase the scars of a conflictual past or to bring it to a harmonious end, these discourses set in motion the belief that the separate histories, geographies, and cultures that have divided humanity are now being brought together by the warm embrace of globalization, understood as a progressive process of planetary integration."[13] This is the ultimate neoliberal fantasy of featureless, ahistorical, and abstract space, and it is an extremely powerful one. Only through a deliberate countering of this set of images and ideas through a deep engagement in actual and imagined people's histories and geographies, and through the painstaking investigations of liberal and nonliberal practices and productions, can this neoliberal fantasy be exposed as the false god it is and laid, at last, to rest.

# Notes

## CHAPTER ONE

1. Coronil, "Towards a Critique of Globalcentrism," p. 351.
2. Fukuyama, *The End of History*.
3. Li Ka-shing bought the land on which the 1986 World Exposition was held for C\$320 million in 1988. At the time, the land comprised nearly one-sixth of the entire Vancouver downtown. The price was considered extremely low even then, promoting speculation that provincial politicians had deliberately marketed the land to a well-known Hong Kong developer in the hopes of stimulating further Asian investment in the region. The deal was roundly criticized in the media. See, e.g., Robert Matas, "Mystery, Unanswered Questions Remain about B.C.'s Land Deal of the Century," *Canada Globe and Mail*, June 17, 1989, A1.
4. By "unprecedented" I am referring to the numbers of people who moved with large capital assets to a specific city in a relatively short period. For a good overview of Hong Kong emigration during this time, see the edited collection by Skeldon, *Reluctant Exiles?*
5. I refer here to the Business Immigration Program, which I discuss in greater detail in Chapter 2.
6. Of course poorer immigrants continued to arrive in Vancouver from Hong Kong as well. In this study I focus on the wealthy precisely because of their ability to affect local structures of power/knowledge in a conspicuous and sustainable way. I use the terms "white" and "Chinese" to differentiate between groups as they self-identify and as they are largely perceived and defined in the societies I am studying. I also use the term "race" as a way of talking about these perceptions and definitions. My use of these terms is not meant to indicate fixed or scientific categories of difference.
7. Hall, "The Problem of Ideology."
8. Mehta, *Liberalism and Empire*, p. 11.
9. Ibid., p. 4.
10. Hall, "The Toad in the Garden," pp. 69–70.
11. I borrow the term "nation format" from Wimmer and Glick Schiller, "Methodological Nationalism and Beyond."
12. Ong, *Flexible Citizenship*. See also Sklair, *The Transnational Capitalist Class*, which focuses on the transnational movements of the wealthy and on the institutional changes set in motion through these types of global flows. Sklair, however, is more concerned with the dominant worldviews and social organization of corporations, and less concerned with the constitution of individual subjects or their movements through space.

13. Ong, *Flexible Citizenship*, p. 195.

14. Ong focuses on the state-led implementation of a liberal economic logic in "Asia" rather than in Western liberal democracies, which does give her authoritarian reading of liberalism more purchase. Nevertheless, the use of a pan-Asian conceptualization remains problematic, given the wide variety of actually existing liberalisms across urban and national regimes in Asia.

15. Appadurai, *Modernity at Large*, esp. p. 177.

16. Sparke, *Hyphen-Nation-States*, chapter 2.

17. See, for example, Brenner, "State Territorial Restructuring"; Smith, "Geography, Difference"; Swyngedouw, "Excluding the Other."

18. Marston, "The Social Construction of Scale."

19. For further discussion of the process of "self-orientalizing," see Ong, "On the Edge of Empires"; and Dirlik, "Chinese History."

20. For an important early study investigating the spatialization of racial hegemony, see Anderson, *Vancouver's Chinatown*.

21. For my critique of Bhabha on this general score, see K. Mitchell, "Different Diasporas."

22. Although I use the term "sedimented" to indicate the spatialization of hegemony through time, I do not mean to indicate that these landscapes are in any way fixed or unchanging. In fact, the premise of this book rests on the notion of hegemony as a process constantly in flux, as are the "landscapes" of hegemony that underpin it.

23. There are hundreds of contemporary books on globalization, but few offer a spatial emphasis. Some of the best recent studies of globalization written from a geographical perspective include: Dicken, *Global Shift*; Cox, *Spaces of Globalization*; Agnew and Corbridge, *Mastering Space*; Olds et al., *Globalisation and the Asia-Pacific*. For a good review of recent globalization texts, see Sparke, "Networking Globalization."

24. Abu-Lughod, *Before European Hegemony*.

25. Beck, *What Is Globalization?* Held et al., *Global Transformations*.

26. Brennan, *Globalization and Its Terrors*.

27. Beck, *What is Globalization?* p. 2.

28. Brennan, *Globalization and Its Terrors*; Katz, "Vagabond Capitalism."

29. Hirst and Thompson, *Globalization in Question*.

30. Thanks to Matt Sparke for his astute observations here. See Sparke, *Introduction to Globalization*.

31. One of the most public battles over the process of globalization vis-à-vis its connections with systems of neoliberal governance was fought in my hometown of Seattle during the meetings of the World Trade Organization (WTO) in 1999. The growing power of the WTO is a classic example of the contemporary neoliberalization of many global institutions. This organization promotes a laissez-faire, free trade rhetoric, preaching rationality and universality, yet its policies are patently skewed in favor of wealthier nations. The failure of poorer nations to win any significant concessions on

agricultural subsidization in the wealthy nations, for example, led to the collapse of the WTO meetings in Cancun in 2003.

32. Held et al., *Global Transformations*, p. 237.

33. Cindi Katz has been one of a handful of geographers to consistently call attention to the crucial role of social reproduction in the constitution of regimes of urban accumulation. See, e.g., Katz, "Hiding the Target." For a discussion that links shifts in social reproduction specifically to new kinds of transnational flows, see Mitchell, Marston, and Katz, "Life's Work."

34. Beck, *What Is Globalization?* p. 4. Italics in the original.

35. Glick Schiller and Fouron, *Georges Woke Up Laughing*, p. 3.

36. The tremendous impact of these remittances for the "home" societies has led to state policies and practices designed to facilitate transnational life for these migrants. See Guarnizo, "Rise of Transnational Social Formations."

37. I am drawing here from the early work of transnational migration theorists, especially Glick Schiller, Basch, and Szanton Blanc, *Towards a Transnational Perspective*. See also their *Nations Unbound*. For other early elaborations of these themes, see Rouse, "Making Sense of Settlement"; and Kearney, "Borders and Boundaries."

38. See, for example, Schiller and Fouron, *Georges Woke Up Laughing*, p. 3.

39. There are now a plethora of these types of studies. For insightful overviews of this work, see Guarnizo and Smith, "The Locations of Transnationalism"; Vertovec, "Conceiving and Researching Transnationalism"; and Portes, Guarnizo, and Landolt, "Introduction."

40. Some exceptions include the work of Aihwa Ong and Leslie Sklair. See also various articles by Beaverstock (e.g., "Transnational Elites") and Yeoh, Huang, and Willis (e.g., "Global Cities").

41. For a discussion of the multiple passport strategy executed by the wealthy and upper middle class in Hong Kong, see Ong, *Flexible Citizenship*, p. 1. For the strategic use of class advantages and cultural capital in evading state disciplinary regimes, see also Hannerz, "Cosmopolitans and Locals."

42. Castells, *Rise of the Network Society*; Gereffi and Korzeniewicz, *Commodity Chains and Global Capitalism*.

43. Smart and Smart, "Transnational Social Networks"; Naughton, "Between China and the World."

44. These "entrepreneurs" were among those targeted by the Business Immigration Program in Canada in the 1980s. Many had a difficult time establishing successful businesses in Canada and either retired early or returned to work primarily in Asia. See Ley, "Seeking 'Homo Economicus' Transnationally"; and Smart, "Business Immigration to Canada."

45. There is, of course, no unitary outlook or reception in neighborhoods or families, or even by individuals. Both individual and group identities always occupy multiple and contradictory positions that overflow these types

of boundaries. My phrasing is meant to identify some general tendencies of a particular moment, rather than to attach specific views to specific groups.

46. Although Antonio Gramsci and Stuart Hall are the obvious exceptions here, they too often adopt an abstract terminology of space and, as discussed later, tend to locate these metaphoric spaces within a nation/territory format.

47. Althusser, "Ideology and Ideological State Apparatuses."

48. Hall, "The Toad in the Garden."

49. Larner, "Neo-Liberalism," p. 13. Italics mine.

50. Hall, "The Toad in the Garden," p. 42.

51. Jessop, "Liberalism, Neoliberalism, and Urban Governance," p. 466.

52. Like Hall, I am less interested in the psychoanalytical examination of the stages of infant development and subjectivity formation as theorized by figures such as Lacan, and much more in the intellectual and emotional changes experienced by adults. See Hall, "The Toad in the Garden," p. 50.

53. Williams, *Marxism and Literature*, pp. 120, 108.

54. Marx and Engels, *The German Ideology*, pp. 42, 43, 64.

55. Marx, *Eighteenth Brumaire of Louis Bonaparte*.

56. Gramsci, *Selections from the Prison Notebooks*, p. 408.

57. Luxembourg's thoughts on this topic are referenced in ibid., p. 404.

58. Neil Brenner and Neil Theodore also use the phrase "actually existing neoliberalism" to argue for embedded analyses and an emphasis on the strategic role of cities. Their overall emphasis on embedding, however, privileges the economic to a point that a Gramscian understanding of the articulation of economic, social, political, and cultural processes in hegemony is completely lost. See Brenner and Theodore, "Cities and Geographies."

59. The "path" taken by twentieth-century liberalism in the West was not in any way given or natural, but rather the result of a number of determining factors, including the nineteenth-century crisis of laissez-faire capitalism and the struggle by a number of social forces to end, or at least temper, its destructive excesses. Social liberalism is thus perhaps best described as a short-term compromise by those who were struggling, at the time, for a more progressive and liberatory solution to capitalism's inherent contradictions. See Polanyi, *The Great Transformation*.

60. See Katznelson for a discussion of the efforts of early-twentieth-century "new" liberals such as Hobhouse and Laski to formulate a socially-based, state-friendly liberalism of this kind. Hobhouse wrote in 1911, for example, "Liberty and compulsion have complementary functions, and the self-governing State is at once the product and the condition of the self-governing individual." See Hobhouse, *Liberalism*, p. 81. Cited in Katznelson, *Liberalism's Crooked Circle*, p. 64.

61. See, for example, the starting point taken by some of the most influential contemporary liberal thinkers in Canada, e.g., Taylor, *Sources of the Self*; and Kymlicka, *Multicultural Citizenship*.

62. The works of Nikolas Rose and Mitchell Dean are important exceptions here.

63. Jessop, "Liberalism, Neoliberalism, and Urban Governance," p. 464.

64. Agnew and Corbridge, *Mastering Space*, chapter 7; Katznelson, *Liberalism's Crooked Circle*, p. 16.

65. The legacy of Britain's structures of liberal governance in the Canadian system of constitution and law are evident in Alan Cairns's discussions of constitutional development and reform in Canada. See, in particular, *Constitution, Government, and Society*. I discuss the "problem" of Québec for Canadian liberalism in Chapter 3.

66. Mehta, *Liberalism and Empire*.

67. Gray, *Liberalism*, p. x.

68. Macpherson, *Political Theory of Possessive Individualism*, p. 221.

69. Pateman, *The Sexual Contract*, p. 53. The internal quotes are from Locke, *Two Treatises of Government*, p. 47. Pateman's critique, which I touch on briefly here, is just one of numerous feminist interventions in the political theory of liberalism. Many of these investigate the problem of political exclusion premised on the separation of the public and private spheres integral to liberal theory. See, e.g., Okin, *Justice, Gender, and the Family*; Elshtain, *Public Man, Private Woman*.

70. Mehta, *Liberalism and Empire*, pp. 57–76.

71. Ibid., pp. 63–64. One of the many justifications for empire was the false promise that through proper tutoring and the right kind of civilizational progress, colonial subjects might eventually aspire to the level of understanding necessary for full inclusion in political society. This promise was false on a number of fronts, but especially in the sense that social distinction, as Bourdieu has indicated, is not just learned through formal systems of education but is acquired through the rhythms and habits of everyday life. The right to inclusion is thus literally *embodied* in the individual and the environment he inhabits. See Bourdieu, *Distinction*.

72. Marshall and Bottomore, *Citizenship and Social Class*. By social citizenship Marshall was referring to a minimum level of material well-being necessary for an individual to be able to participate in political society. In this view, adequate food, housing, and education are fundamental rights for all citizens, without which the citizen is effectively disenfranchised.

73. Ibid.

74. To get a flavor of the U.S.-centrism of Dewey's work, see *Dewey, The Later Works*, vol. 11, pp. 235 and 357. For a critique of Dewey's profound Americanism, see K. Mitchell, "Education for Democratic Citizenship."

75. Hobhouse argues that even the later writings of John Stuart Mill showed a revisionist tendency against the negative conception of freedom as noninterference, and toward a greater acceptance of "effective freedom or freedom as ability" stemming from Hegelian philosophy. The positive view of freedom led many to a "defense of enhanced governmental activity and authority and to support for measures limiting contractual liberty"—thus undermining a key foundation of Lockean liberalism. See Gray, *Liberalism*, p. 32.

76. Katznelson, *Liberalism's Crooked Circle*, p. 65.

77. The most famous early elaboration of these principles was Rawls, *A Theory of Justice.*

78. Mehta, *Liberalism and Empire,* p. 20.

79. Mulhall and Swift, *Liberals and Communitarians,* p. xvi.

80. Nozick, *Anarchy, State, and Utopia.*

81. For an interesting overview of the rise of welfarist principles of governance and its connection with democratic development in numerous modernizing states, see the essays in Marshall and Chapman, *The Social Construction of Democracy.*

82. With respect to the primary role of consent as the "ideological lynchpin of western capitalism," see P. Anderson, "The Antinomies of Antonio Gramsci."

83. Gray writes of nineteenth-century liberal revisionism; for example: "It is important to note that the decline of classical liberalism cannot be explained simply by a response to the abandonment of important classical liberal ideas by John Stuart Mill and others. Such developments in intellectual life reflected, and were in part caused by, the changes in the political environment brought about by the expansion of democratic institutions." Gray, *Liberalism,* p. 33.

84. See Katznelson, *Liberalism's Crooked Circle,* for a contemporary theoretical position along these lines.

85. See, for example, the essays in Neil Nevitte, *Value Change and Governance in Canada* (Toronto: University of Toronto Press, 2001).

86. Some of the other liberal keywords or concepts that were referred to in popular discourse and that appear throughout this book are "universality," "choice," "civilization," "progress," and "consent." As I hope to show, these concepts rely on implicit processes and understandings of differentiation, hierarchy, and normality that serve ultimately to limit universal social and political membership in the society.

87. Steven Lukes writes: "Liberalism is about fairness between conflicting moral and religious positions, but it is also about filtering out those that are incompatible with a liberal order and taming those that remain." See Lukes, *Moral Conflict and Politics,* p. 18, n. 39.

88. Mehta, *Liberalism and Empire,* p. 20.

89. It is important to note that although these proto-neoliberal interests (critiquing the tenets and practices of social liberalism) coalesced at a particular moment here, they did so from different vantage points and for different reasons.

90. Peck and Tickell, "Neoliberalizing Space."

91. For example, Donzelot's phrase *"Formation permanente"* indicates a state of being in continual retraining or constant mobilization in the domain of work. According to Donzelot, this way of being in the world is becoming normalized within the contemporary neoliberal regime and represents a profound shift vis-à-vis the relation between the subject and society. See Donzelot, "Pleasure in Work."

92. Rose, "Governing 'Advanced' Liberal Democracies."

93. Rose, *Governing the Soul*, p. 264.

94. Larner, "Neo-Liberalism," p. 14. In making this point, Larner has drawn on O'Malley, "Indigenous Governance," and Frankel, "Confronting Neo-Liberal Regimes."

95. Peck and Tickell, "Neoliberalizing Space," p. 393.

96. The phrase "a shout in the street" is taken from Berman, *All That Is Solid*.

97. For an in-depth examination of one of the main urban megaprojects in Vancouver's downtown, the Concord Pacific development by Li Ka-shing, see Olds, *Globalization and Urban Change*.

98. Smith, "New Globalism, New Urbanism," p. 428.

99. Ibid., p. 443.

100. Taylor, "The Politics of Recognition."

101. See, for example, the essays in George, *Burning Down the House*.

## Chapter Two

*Epigraph.* Keil, "Common-Sense Neoliberalism," p. 579.

1. Smith, *The New Urban Frontier*.

2. Smith, "Giuliani Time." MacLeod reiterates that the original referent was the "right-wing 'revanchist' populist movement, which throughout the last three decades of the 19th century reacted violently against the relative liberalism of the Second Empire and the socialism of the Paris Commune." MacLeod, "From Urban Entrepreneurialism," p. 606.

3. Ley, "Styles of the Times," p. 46. See also Ley, "Liberal Ideology."

4. See, e.g., Walker and Greenberg, "Post-Industrialism and Political Reform.

5. For a comparative case in Ontario, see Keil, "Common-Sense Neoliberalism," p. 592.

6. City of Vancouver, "Apartment Vacancy Rates, 1976–1992."

7. Shelly Easton, "City's 0.3% Vacancy Rate So Small It's Barely There," *Vancouver Province*, November 26, 1989.

8. Vacancy rates in 1989 for studios were 0.4 percent, and for three-bedroom apartments 0.9 percent. In 1990, vacancy rates for studios were 0.7 percent but were 2.0 percent for three-bedroom apartments. See Canada Mortgage and Housing Corporation, "Vacancy Rates."

9. Ann Rees, "Desperate Valley Renters Plead for Help," *Vancouver Province*, May 24, 1990.

10. "Rentals High and Low," *Vancouver Province*, September 3, 1989.

11. Average rents in the West End leaped from C$500 to C$750 in a single year. See Robert Sarti, "West End Landlords Cashing in on Rental Squeeze, Tenants Charge," *Vancouver Sun*, November 15, 1989; see also Robert Sarti and Susan Balcom, "Rental Crisis: Rising Rent, Zero Vacancy Put Squeeze on Tenants," *Vancouver Sun*, November 30, 1988; Katherine

Monk, "Vancouver-Area Vacancy Rate Still National Low, Survey Shows," *Vancouver Sun*, November 24, 1990; and Anne Gregory, "Rent Increases Making Life Difficult," *Vancouver West Ender*, October 25, 1990.

12. See Jean Kavanagh, "Sex Harassment in Rent Crisis Charged," *Vancouver Sun*, November 22, 1989; "Rental Picture a Disaster," *Vancouver Province*, November 3, 1989; and "No Place to Go," brochure, Tenants' Rights Coalition, 1989.

13. The idea of rescaling the urban imaginary comes from Bob Jessop, "A Neo-Gramscian Approach."

14. Knox, "Capital, Material Culture," p. 7.

15. City of Vancouver, Planning Department, *False Creek: A Preliminary Report from the False Creek Review Panels*, September 1973.

16. Zukin, *Loft Living*, p. 175.

17. The Canadian Pacific Railroad used the land for transportation purposes after 1892. By the 1920s the shoreline was completely industrialized, with 5.5 miles of shipyards, sawmills, and metal factories. See Cybriwsky, Ley, and Western, "Political and Social Construction," p. 108.

18. Marathon refused to develop the land because of the saturation of the luxury housing market in the 1970s and the city's insistence (during the TEAM years) on lower densities and on the provision of social housing for low-income residents. See Gutstein, "Expo's Impact on the City;" Cybriwsky, Ley, and Western, "Political and Social Construction," p. 111.

19. See Ley, "Styles of the Times," for a good discussion of the early development of the north shore of False Creek from 1980 to 1986, prior to its configuration as the Expo lands.

20. Anderson and Wachtel, *The Expo Story*, p. 13.

21. Ibid. The package included twenty-six bills that attacked working conditions, unionization, and rent control, among other things.

22. Gutstein, "Expo's Impact on the City," p. 73.

23. Quoted in Gutstein, "Expo's Impact on the City," p. 66.

24. Other major projects included Canada Harbour Place, an international trade and convention center, massive highway building and repair (the highway budget was increased by C$110 million in 1986), and the C$60 million Cambie Street Bridge.

25. Anderson and Wachtel, *The Expo Story*, p. 24.

26. One downtown resident said at the microphone: "I've been evicted, Mr. Madden, because my rent is going up in order to clear us out and make our hotel a bed and breakfast place. Our homes are hotels, do you understand that?" Quoted in ibid., pp. 5–6.

27. Pred, "Spectacular Articulations of Modernity," p. 50.

28. Mark Hume, "Expo Land Firms Woven into Complex Crown Web," *Vancouver Sun*, April 8, 1988.

29. "B.C. Expo Lands: Sale of the Century," *Toronto Globe and Mail*, June 17, 1989, A7.

30. Robert Matas, "Mystery, Unanswered Questions Remain about B.C.'s Land Deal of the Century," *Toronto Globe and Mail*, June 17, 1989, A1.

31. Author's interview in Hong Kong, May 4, 1991. With the exception of public figures, authors, and reporters, I have used pseudonyms to preserve confidentiality. All interviews took place in Vancouver, unless otherwise noted.

32. Tom Tevlin, "'Twilight' Zoning? Government Zoning Letter Has City Fuming," *Vancouver Sun*, May 8, 1988; see also, Sarah Cox, "City Shuns Expo Site Payoff," *Vancouver Sun*, May 11, 1988.

33. Gordon Hamilton, "Victoria Offered to Change Law over Expo Sale, Contract Shows," *Vancouver Sun*, April 11, 1989.

34. Ibid.

35. Gordon Hamilton, "Expo Site Purchaser Identified," *Vancouver Sun*, February 17, 1990.

36. The government normally pays C$22 per buildable square foot for social housing. See Robert Matas, "Expo Land for Sale at Triple Original Price," *Vancouver Sun*, August 30, 1989, A1.

37. Gordon Gibson, "Let's Buy the Whole Mess Back," *Vancouver Sun*, January 26, 1990, A15.

38. From a press statement by John Chan, "Press Conference for the Canada Festival in Hong Kong 1991," April 30, 1991.

39. Author's interview with the executive secretary of the Canadian Chamber of Commerce, Hong Kong, May, 1991.

40. The Canadian government paid 22 percent of the costs of the festival.

41. These included the Hongkong and Shanghai Banking Corporation Limited (Hongkong Bank), the Canadian Imperial Bank of Commerce, the Bank of Montreal, the Royal Bank of Canada, the Royal Canadian Mint, and the Bank of Nova Scotia.

42. Harvey, "Flexible Accumulation through Urbanization," p. 280.

43. Prime Minister Mulroney quoted in *Canada and Hong Kong Update*, Spring 1991.

44. "15 Million Expected to Visit Expo '86," *Hong Kong Standard*, October 17, 1985.

45. "Window on the Pacific."

46. Statistics Canada, *Employment and Immigration Canada, 1985*, p. 1.

47. Quoted in the *Financial Post*, December 13, 1989, p. 7.

48. In 1991 this was amended so that investors seeking to move to provinces that had already received a significant number of business immigrants (British Columbia, Ontario, and Quebec) would be required to invest C$350,000 for five years.

49. The amount of investment required varies by province and has been raised C$100,000 since late 1990. British Columbia is a Tier 2 or "have" province, which requires a minimum investment of C$350,000 (formerly C$250,000). Tier 1 or "have not" provinces require C$250,000 over three years. These provinces are Newfoundland, Nova Scotia, New Brunswick, Prince Edward Island, Manitoba, and Saskatchewan. See Employment and Immigration Canada, *Immigration Regulations, Guidelines, and Procedures*;

and "Visa for Dollars Plan Hits Trouble," *Financial Post Daily* (Canada), May 21, 1991, p. 1.

50. Statistics Canada, *Immigrants in Canada*, p. 30.

51. Employment and Immigration Canada, *Doing Business in Canada*. See also Smart, "Business Immigration to Canada; Nash, *The Economic Impact*; and Yeung, "Hong Kong's Business Future."

52. In Hong Kong, the ability to jump the processing queue could mean the difference of several years in waiting time. Applicants were assessed for entry into Canada based on a 1967 point system: entrepreneurs got an extra twenty points in lieu of being assessed under the usual categories of occupational demand and occupational skill. See Nash, *The Economic Impact*, p. 7.

53. Employment and Immigration Canada, *Immigration to Canada*, p. 41; Lary, "Trends in Immigration," p. 6.

54. Lary, "Trends in Immigration."

55. These statistics are from the Commission for Canada Immigration Section, Hong Kong.

56. In these statistics, 809 visas were issued under the entrepreneur program to business immigrants destined for British Columbia in 1990, and 436 (49 percent) visas were issued under the investor program; see "British Columbia Immigration Highlights," *BC STATS 90*, 4 (May 1991). It is worth noting that statistics on immigration vary considerably depending on the source. The figures offered here are intended to show general patterns.

57. Lary, "Regional Variations," p. 6.

58. Employment and Immigration Canada, *Annual Immigration Statistics*.

59. Ley, "Seeking 'Homo Economicus' Transnationally."

60. Aihwa Ong has also come across this term in her fieldwork, indicating its widespread use. See Ong, "On the Edge of Empires."

61. For a further discussion of the FTA and its links with neoliberalism, see Clarkson, "Constitutionalizing the Canadian-American Relationship"; Drache, "The Future of Trading Blocs." For more on the charter, see Mandel, *The Charter of Rights*. On the constitution, a useful source is McBride and Shields, *Dismantling a Nation*. For an overall analysis of the articulation of these dynamics in Canada's unfolding constitutional debates, see Sparke, *Hyphen-Nation-States*.

62. Jenson, "'Different' but Not 'Exceptional.'"

63. Rekart, *Public Funds, Private Provision*; Jenson, "'Different' but Not 'Exceptional.'"

64. Rekart, *Public Funds, Private Provision*, p. 14.

65. Ibid., pp. 15–16.

66. Smith, "Gentrification, the Frontier," p. 21.

67. Nash, "The Emigration of Business People," p. 3. For comparative figures, see also Phil Macdonald, "Canada to Trim Numbers of Independent Class Visas," *Hong Kong Standard*, January 26, 1990.

68. As Robert Puddester, the head of Canadian immigration in Hong Kong, said in a newspaper interview: "The more money they have, the less

they declare." Quoted in Ben Tierney, "B.C. Now Top Destination for Hong Kong, Taiwan Investors," *Vancouver Sun,* February 11, 1992.

69. My estimate is still far lower than one given by Hong Kong economist and businessman Simon Murray, who believed that Hong Kong lost HK$2.25 billion per month to Vancouver in 1989—nearly C$4 billion per year. See Fanny Wong, "Confidence Crisis Costing Billions," *Hong Kong Standard,* September 21, 1989.

70. Quoted in Marcus Brauchli, "First Generation Starts Up Fund Targeted at Immigrants to Canada," *Asian Wall Street Journal,* October 8, 1986. Within two years, the Maple Leaf Fund amassed a total of C$18 million, involving seventy-two immigrant investors. By 1992, four Maple Leaf investment funds were worth more than C$100 million. The Maple Leaf Fund in British Columbia alone amassed C$45 million, and shareholders voted to wind it up when it matured in 1992. (Owing to the fund's success, the 215 shareholders of the C$32 million Maple Leaf Fund in Manitoba voted to list shares on the Toronto Stock Exchange rather than wind up the fund upon maturation in the same year.) Most of the shareholders of the Maple Leaf funds were originally from Hong Kong and Taiwan and had invested C$250,000 in the immigration-related package three years earlier. Over half had moved to Canada by 1992. See Robert Williamson, "Funk Aspires to Banking," *Toronto Globe and Mail,* May 9, 1992, B2.

71. In 1986, the bank offered three investment vehicles at different prices for shareholders. The first intended to raise C$12 million from four investors at C$3 million each; the second was a C$2.5 million fund with ten equal shareholders; the third was a C$1 million fund with four partners. For each fund, the CIBC acted as escrow and distribution agent, with a 3 percent front-end load plus legal fees. See Paul Baran, "Investor Immigrants Wooed," *South China Morning Post* (Hong Kong), January 14, 1986. CIBC, which bills itself as the "Gateway to Canada" in its promotional literature, was the first bank to establish investment funds that targeted Hong Kong emigrants. According to many sources, these funds produced very low (3 percent) returns on the early investments.

72. First Canadian Bank, Bank of Montreal, "Branches and Offices Serving the Chinese Community," brochure, Hong Kong, 1991. The brochure is in both English and Chinese. In February 1991, the Bank of Montreal's mortgage-backed private placement immigration fund required an investment of C$250,000 and promised to pay 7 or 8 percent annual interest, maturing in three years.

73. Setting up offshore trusts became legal in Canada after 1972.

74. Quoted in William Claiborne, "Vancouver: Another Little Dragon in the Making," *International Herald Tribune,* May 14, 1991.

75. The Hong Kong Bank of Canada was founded in Vancouver in 1981 as a small branch of the international giant. In 1986 the bank acquired the assets of the Bank of British Columbia for C$63.5 million, and in 1990 it bought out Lloyds Canada. By 1991, the bank had the largest consumer presence of any foreign-owned bank in Canada and had opened 107 Canadian

offices, including 16 that specialized in Asian banking. Of its retail deposits (over $C5 billion), 25 percent originated from Asian customers. *Toronto Globe and Mail*, August 5, 1991, p. B3.

76. Jim Lawrie, vice president and Hong Kong–based private banking chief for Asia Pacific. Quoted in Eva To, "Canadian Bank Eyes Emigrants from HK," *Business News*, May 12, 1990.

77. Royal Bank of Canada, "Emigrating to Canada?" brochure, Hong Kong, 1991.

78. The service package was offered in March 1988. See Cynthia Suen, "Canadian Bank Sets Sights on New Immigrants," *South China Morning Post* (Hong Kong), October 24, 1988.

79. Rich Gossen, "Double Benefit for Emigrants," *South China Morning Post* (Hong Kong), July 1, 1990.

80. See Knox, "Capital, Material Culture"; Logan, "Cycles and Trends"; Olds, *Globalization and Urban Change.*

81. Tickell, "Global Rhetorics, National Politics," p. 155.

82. Knox, "Capital, Material Culture," p. 5.

83. Olds, "Globalization."

84. Dalton, "Canada," p. 17.

85. Victor Fung, "Hong Kong Investment Funds Pour into Canada," *Financial Post* (Canada), June 17, 1991, p. 18.

86. Lawrence Lim, quoted in Moira Farrow, "Hong Kong Capital Flows Here Ever Faster," *Vancouver Sun*, March 21, 1989.

87. In several of the major Vancouver-based investment syndicates operative in 1991, developmental real estate projects figured prominently. For example, the financial activity of President Canada Syndicates Inc., located in Burnaby, showed an investment of C$5,379,460 by way of a shareholder's loan in a wholly owned subsidiary called President Asian Enterprises Inc. Between 1989 and 1991, this subsidiary had invested in six developmental real estate properties in the greater Vancouver area. Investment in the syndicate subsidized the development of these properties, including a mammoth hotel–shopping plaza in Richmond. ("Offering Memorandum," p. 5; available from President Canada Syndicates Inc.) Other Vancouver investment funds that responded to my information request included Canada New Life Fund Ltd., Vancouvplus Capital Corporation, and Golden Beaver Development Corp. All were involved in ongoing real estate development projects in Vancouver. Funk's Maple Leaf Fund investments were similarly involved in a number of developments, including the C$581,000 investment in the Blueberry Hill residential-commercial development at Whistler. See Gutstein, *The New Landlords*, p. 96.

88. For example, Perfect Coins International was a subsidiary of Perfect Coins Investment Group, a firm originally involved in foreign exchange. The subsidiary company operated purely in the property market, concentrating mainly on the Hong Kong–Vancouver connection.

89. Other consultants described even greater services provided for their Hong Kong clients. ACCESS Canada, established in 1988, provided infor-

mation and photographs of houses and house prices in Canadian cities, information about schools, and a data bank of businesses for sale. Vigers Hong Kong offered the computer system VICCI, open twenty-four hours a day and linked up with all overseas property offices, according to my interview with a Vigers agent, Hong Kong, August 1991.

90. I discuss the redevelopment of the west-side neighborhoods at greater length in Chapter 5.

91. Donald Gutstein, an investigative journalist who examined Vancouver real estate in the 1980s, estimated in our 1990 interview that about half the developers involved in the Kerrisdale "condo-mania" were Asian. In terms of financing, Michael Geller, of the Urban Development Institute of Canada (a developers' lobby group) estimated that approximately 60 percent of the 3,000 condominiums planned for construction in the greater Vancouver area in 1989 were financed in part or completely by Asian investors. See Gutstein, *The New Landlords*, p. 167. For individual case histories of specific apartment blocks in Kerrisdale, see also Daphne Bramham, "Condo Craze Creates Risks," *Vancouver Sun*, August 18, 1989, B1. Differences in estimates and controversies over the actual amount of Hong Kong investment in Vancouver real estate reflect not only the difficulties of tracking capital flows, but also the attempt by various parties to manipulate these statistics for specific ends.

92. These Hong Kong companies were also involved in joint ventures with local developers such as Polygon, United Properties, and Buron and Molnar, who solicited or distributed projects through wholesalers such as Park Georgia Realty, Hong Kong Midland Realty, L & D Associates or Royal LePage Asian Affiliates. All these developers were headquartered or had offices and extensive networks operating out of Hong Kong.

93. Quoted in a Colliers Macaulay Nicolls informational letter for Kerrisdale Towers, August 10, 1989. The ad went on to recommend twin towers "recently renovated to reflect the high expectations of the tenants." See also the advertisement in *West Side Weekly*, November 9, 1990, for the new luxury condos at Connaught Place, 2128 West Forty-third. The photograph of the building is captioned, "A Gracious Way of Life." Many of the condos have perks such as putting greens, outdoor hot tubs, indoor saunas, and Italian marble foyers. See, for example, the Royal LePage informational brochure on Kerrisdale Parc Condos.

94. Royal LePage informational brochure, November 1990.

95. From *Vancouver West Side Real Estate*, November 3, 1989, p. 87; and *Vancouver Real Estate Weekly*, October 27, 1989.

96. See "Firm Real Estate Specialists," *Vancouver Courier*, August 23, 1989.

97. See, e.g., Sarah Cox, "Buying Up Their New World," *Vancouver Sun*, February 27, 1988; Keast, "The World's Longest Commute"; "Resentment Feared as Condos Sold to Hong Kong Residents," *Vancouver Sun*, December 14, 1988.

98. The practice of flipping was featured in other articles around this time, e.g., "Flippers Awash in Profits," *Vancouver Sun*, February 8, 1989;

"City Housing Market Flipping Along," *Vancouver Business*, September 9, 1989.

99. Gillian Shaw, "Money Is King in Hong Kong: Entrepreneurs Find Paradise in the Streets of Hong Kong," *Vancouver Sun*, March 21, 1989, A1.

100. David Lam, who immigrated from Hong Kong in the late 1960s and became a multimillionaire through real estate investment in Vancouver, was appointed lieutenant governor of British Columbia in 1988. See Wiseman, "On Guard for Thee," p. 130.

101. Quoted in Carl Nolte, "Racial Tension on Rise in Vancouver," *English Newspaper* (Hong Kong), July 9, 1989.

102. Hong Harbour was the nickname for the upmarket False Creek area; Hongkong Row was the name for a section of Oakridge that was popular with Chinese investors.

103. Jerry Collins, "Vancouver Feels the Asian Strain," *South China Morning Post* (Hong Kong), March 4, 1989. See also, "Paying for Innocence," *South China Morning Post* (Hong Kong), January 14, 1989.

104. Some signs were even more explicit. "I am disposable" said one. See *Vancouver Sun*, August 15, 1989, B1. Another, held aloft by an elderly women wearing an orange Caution sign over her shirt, said, "Taxes Up Business Down Rents Up Landlords Gouge (25–35%) Good Housing Demolished Where Do We Go?" From a photo in Noreen Shanahan, "Elderly Women Hit Hard," *Vancouver Western News*, July 1989.

105. See Daphne Bramham, "Bulldozer Razes Seventh Apartment," *Vancouver Sun*, October 24, 1989.

106. Quoted in Glen Schaefer, "Protesters Got a Message across to the Developers," *Vancouver Province*, January 26, 1990. (Note: The name "Annie Humphreys" is a pseudonym.)

107. City of Vancouver, Planning Department, Permits and Licenses Department, "Demolitions" file, 1992.

108. For example, in one Kerrisdale area in 1989, 116 rental units were destroyed to make way for 67 luxury condominiums, and 359 units were pending destruction to make way for 222 condominiums.

109. The West End should not be confused with west-side neighborhoods. The West End is downtown, adjacent to Stanley Park, whereas west-side neighborhoods are adjacent to the University Endowment Lands.

110. The phrase is from Katz, "Vagabond Capitalism."

111. Some examples: Shanahan, "Elderly Women Hit Hard"; Daphne Bramham, "Radicals are Bred amid Turmoil," *Vancouver Sun*, August 17, 1989; Donna Anderson, "Bulldozer Blues," *Vancouver Sun*, December 15, 1989; Bramham, "Bulldozer Razes Seventh Apartment."

112. Frances Bula, "Kerrisdale Land Flips Spark Anger," *Vancouver Sun*, February 26, 1989. People I interviewed who had been present at the early meetings mentioned strong anti–Hong Kong reactions.

113. "Angry Residents Delay Kerrisdale Demolition," *Vancouver Sun*, August 18, 1989; "Apartment Demolitions Protested," *Vancouver Sun*,

August 15, 1989; "Mayor Blamed for Housing Crisis," *Vancouver Province,*
August 15, 1989.

114. "RSVP Wants Action," *Vancouver Courier,* April 5, 1989.

115. Mehta, *Liberalism and Empire,* p. 117.

116. Williams, *The Country and the City.*

## CHAPTER THREE

1. Cairns, *Constitution, Government, and Society,* p. 91. The 1840 Act
of Union, a response to Lord Durham's negative assessment of French
Canada in a report of 1839, was an effort to bring the English and French
together in a unitary state dominated by Anglo-Saxon cultural and economic
mores.

2. Before the British North America Act, both the Québec Act of 1774
and the Constitution Act of 1791 guaranteed French Canadians full equality with the English-speaking regions of Canada. See Parekh, *Rethinking
Multiculturalism,* p. 185.

3. For an interesting popular critique of some "core" philosophical differences between the United States and Canada, see Lipsett, *Continental
Divide.* These include the Canadian perception of the United States as
ultraindividualist and -nationalist, with an overweening emphasis on assimilating (immigrant) difference into an American national identity.

4. However, as I will show with the Meet with Success program, this
"international" cosmopolitanism remained firmly rooted in the individualist
values of a secular British liberalism.

5. Rouse, "Questions of Identity, p. 356.

6. Macpherson, *Political Theory of Possessive Individualism.*

7. Fei, *From the Soil.*

8. Rouse, "Questions of Identity," p. 368.

9. Ibid.

10. For a general critique of cultural differentiation of this sort, see Benhabib, *The Claims of Culture.*

11. Theorizing cultural difference within the territorial container of the
nation is common to both liberal and communitarian thinkers. Some are
explicit in making this connection; see, e.g., Kymlicka, *Multicultural Citizenship,* p. 93. Others rely more implicitly on the territorial boundaries of
the nation as the glue binding different spheres of justice; see, e.g., Walzer,
*Spheres of Justice.* William Connolly writes critically of this territorial
bounding: "When it comes to the wall dividing democratic politics inside
the state from the outside, Ricoeur joins Rousseau, Schumpeter, and Walzer
in overcoding the boundary of the territorial state" (*The Ethos of Pluralization,* pp. 148–149).

12. For other works that discuss the connections between multiculturalism and capitalism, see Smith, "What Happened to Class?"; K. Mitchell,
"Multiculturalism"; and Zizek, "Multiculturalism."

13. Compare Yasmin Alibhai-Brown, *After Multiculturalism*.

14. The Meech Lake Accord was a proposed constitutional amendment that recognized French culture and presence as a "fundamental characteristic" of Canada. The accord included a "distinct society" clause that would allow Québec special rights within the framework of the Canadian state. It was not ratified by the provincial legislatures in time in 1989, and thus failed to become law. Numerous commentators have reflected on the fundamentally British, individualist, and proceduralist form of liberalism of the Canadian Charter of Rights. For a discussion of this, and related points associated with the "problem" of Québec and contemporary liberalism in Canada, see Taylor, *Reconciling the Solitudes*; Taylor, "The Politics of Recognition"; Webber, *Reimagining Canada*; and Kymlicka, *Finding Our Way*.

15. Zizek, "Multiculturalism," p. 44.

16. See, e.g., Parekh, *Rethinking Multiculturalism*, chapter 6, for a discussion of the liberal state's inability to tolerate "deep diversity."

17. The neglect of indigenous voices in this struggle manifests the general silencing of First Nations people during this period.

18. Laurier's interest in promoting cultural pluralism was aimed at the two charter groups and not intended to include indigenous people.

19. Canada's expansion westward was greatly aided by the National Policy of 1879, which declared Canada's independence from the westward expansion of the United States and allowed central Canada to rule the western hinterland.

20. From a letter to W. D. Gregory, November 19, 1909, the Laurier Papers, Public Archives of Canada. Quoted in Eddy and Schreuder, "Canadians, Canadiens, and Colonial Nationalism," p. 163.

21. Creighton, *Canada's First Century*, p. 108.

22. Myers, *History of Canadian Wealth*, pp. 265–266.

23. Newman, *The Canadian Establishment*; Olsen, *The State Elite*; Porter, "On Multiculturalism"; see also Porter, *The Vertical Mosaic*.

24. Clement, *The Canadian Corporate Elite*, p. 71.

25. The French and the British were considered different races at this time.

26. Eddy and Schreuder, "Canadians, Canadiens and Colonial Nationalism," p. 168.

27. Ibid.

28. K. Mitchell, "Educating the National Citizen."

29. Wardhaugh, *Language and Nationhood*, p. 36.

30. Behiels, *Prelude to Québec's Quiet Revolution*.

31. Rubinoff, "Multiculturalism," p. 122. Gupta and Ferguson, "Beyond 'Culture,'" p. 7. See also Kymlicka, *Multicultural Citizenship*, p. 93.

32. The report, including the addendum, can be found in *Multiculturalism and the Government of Canada*, p. 10.

33. Pierre Trudeau, from a speech given in the House of Commons Debates, October 8, 1971. This speech can be found at http://collections.ic .gc.ca/albertans/speeches/trudeau.html (accessed February 23, 2004).

34. See, e.g., the quotes in Asad, "Multiculturalism and British Identity," 459.

35. The phrases are from Minister John Patten's news release from the Home Office, "On Being British," July 18, 1989. Quoted in ibid.

36. Cited in *Multiculturalism and the Government of Canada*, p. 46.

37. Kallen, "Multiculturalism," 53.

38. Bullivant, *Pluralism.*

39. Kobayashi, "Multiculturalism."

40. Porter, "On Multiculturalism," p. 69.

41. Bullivant, *Pluralism*, p. 105.

42. Kallen, "Multiculturalism," p. 53. Italics in the original.

43. Li, "Race and Ethnicity," p. 15.

44. See, e.g., Robert Bourassa, "Objections to Multiculturalism," *Le Devoir* (Quebec), November 17, 1971; Taylor, "The Politics of Recognition."

45. Parekh, *Rethinking Multiculturalism*, p. 187.

46. Kallen, "Multiculturalism," p. 52.

47. "Multiculturalism and Policing," p. 42.

48. Asad, "Multiculturalism and British Identity," p. 465.

49. Elliot and Fleras, "Immigration." See also Taylor, *Reconciling the Solitudes.*

50. Wardhaugh, *Language and Nationhood*, p. 201; Kallen, "Multiculturalism," p. 55.

51. See Alibhai-Brown, *After Multiculturalism*, with respect to a similar shift in England during this time.

52. See, for example, the figures from "Multiculturalism Expenditures under Three Federal Programs, 1984–85 to 1990–91," in *New Faces in the Crowd*, p. 33.

53. *The Canadian Multiculturalism Act.*

54. *Multiculturalism . . . Being Canadian*, p. 3.

55. Ibid., pp. 1–2.

56. Quoted in Elliott and Fleras, "Immigration," p. 67.

57. *New Faces in the Crowd.*

58. *Working Together.*

59. "Face of Vancouver to be Radically Altered," *Toronto Globe and Mail*, February 20, 1989.

60. "Is Vancouver Trading Furs for Beads?" *Toronto Globe and Mail*, March 1, 1989.

61. David Baxter, "Population and Housing in Metropolitan Vancouver: Changing Patterns of Demographics and Demand." Vancouver: Laurier Institute/Canadian Real Estate Research Bureau/Bureau of Applied Research, 1989.

62. Ibid.

63. "Study Says Baby Boomers behind Home Price Surge," *Toronto Globe and Mail*, November 16, 1989.

64. "Burb Buyers Hit with Levies," *Vancouver Real Estate Weekly*, April 6, 1990.

65. These donors were Lieutenant-Governor Dr. and Mrs. David C. Lam, Asa Johal, Dr. Peter Lee, Milton K. Wong, the Chan Foundation, the Bank of Montreal, and the PCI Group.

66. Letter to potential members, the Laurier Institute, November 2, 1991.

67. Cruz, "From Farce to Tragedy."

68. James Woodsworth, *Strangers within Our Gates*, p. 234. This influential book was written in 1909, during Laurier's tenure in office.

69. Hannerz, "Cosmopolitans and Locals," p. 245.

70. The immigration and investment seminars were given at the World Trade Centre in Vancouver from January through June 1991. The seminar I attended focused on the Business Immigration Program. Although these seminars were designed to facilitate business linkages, the Being Canadian acculturation video was shown at the beginning of every meeting.

71. The narration is in Cantonese with English subtitles.

72. Monster houses were large houses built primarily in Vancouver's west-side neighborhoods in the late 1980s. They were marketed to and purchased primarily by immigrants from Hong Kong and Taiwan and became extremely controversial because of their alteration of the landscape. I examine this controversy in detail in Chapter 5.

73. It is important to note that the west-side neighborhoods where these house controversies were greatest, and toward which the Meet with Success program was targeted, comprise the wealthiest area of the city. With respect to political and economic clout, these neighborhoods have consistently been the most powerful in Vancouver.

74. Many of the multicultural policies in Canadian history have been criticized as superficial efforts to promote cultural tolerance without any real advance in economic parity or reform; see, e.g., Wardhaugh, *Language and Nationhood*, p. 201; Kallen, "Multiculturalism," p. 55.

75. The Canadian Club program was ostensibly directed toward all Hong Kong emigrants to Canada. Most of the very wealthy, however, had already received Canadian citizenship or visas by the time the program was initiated in 1990. In my two visits to Meet with Success, it appeared to me from the questions asked and the apparel of the respondents that they were not of Hong Kong's social top tier. I was not permitted to interview members of the audience.

76. See also Chakrabarty, "Modernity and Ethnicity in India"; and Gordon and Newfield, "Introduction."

77. Taylor has written extensively on the first two differences as they have played out in Canadian politics, differences which he relates to the failure of the Meech Lake Accord and the ongoing conflict between Québec and the rest of the country. See Taylor, "The Politics of Recognition." For an accessible discussion of the inherent contradictions within multiculturalism, see Baumann, *The Multicultural Riddle*.

78. The Canadian Charter of Rights was legislated in the Canadian Constitution Act of 1982. It was generally patterned after the U.S. Constitution and Bill of Rights.

79. Taylor, "The Politics of Recognition," p. 60.

80. This is, of course, not true of all migrants, but many of the people I interviewed in Hong Kong spoke of their impending moves to Vancouver as conditioned by their social responsibilities as parents, husbands, or children to an extended family network. In this vein, see also Hamilton, "Introduction."

81. See, e.g., K. Mitchell, "Flexible Circulation"; Mitchell and Olds, "Chinese Business Networks"; Hamilton and Chen, "Introduction"; Nonini and Ong, "Chinese Transnationalism."

82. Mehta, *Liberalism and Empire*; see also Singh, "Culture/Wars."

83. Taylor, "The Politics of Recognition," p. 66.

84. Taylor, quoted in Bhabha, "Culture's in Between," p. 33.

85. Bhabha, "Culture's in Between," p. 32.

86. Ibid., p. 33.

87. For a longer critique of Bhabha's elision of spatial production, see K. Mitchell, "Different Diasporas."

88. Asad, "Multiculturalism and British Identity."

89. See, e.g., Appiah's critique of Taylor in "Identity, Authenticity, Survival."

90. Coronil, "Towards a Critique of Globalcentrism," p. 351.

## Chapter Four

1. Ley, "Styles of the Times."

2. The phrase "precarious experiment" is taken from an interview with a Vancouverite and Canadian nationalist who repeatedly impressed on me the vulnerability of both his city and the nation. Author's interview, Vancouver, June 1991.

3. These include the False Creek South Shore housing development, discussed in Chapter 2.

4. In Chapter 2 I discussed the material ramifications of these changes in terms of the privatization, deregulation, rezoning, and international marketing of land in Vancouver.

5. This quote is from a draft of the proposal to create a new zoning bylaw schedule, submitted to City Council on February 15, 1990. This draft was given to me by John Pitts.

6. The City Planning Department attempted to control house size and siting with three RS-1 zoning amendments in 1986, 1988, and 1990. Most involved changing the shape of the house envelope to discourage bulkiness. Neighborhood activists I spoke with felt that the city's downzoning efforts were not tailored effectively to the problems of individual neighborhoods.

7. Quoted in Jeff Lee, "Rezoning Idea Has a Permanent Air," *Vancouver Sun*, May 16, 1990; see also "Shaughnessy Gets Zoning Law to Control Size of New Houses," *Vancouver Sun*, June 15, 1990.

8. Davis, *City of Quartz*.

9. "The Neighbourhood Green," p. 1.

10. This brought their total to six out of twelve, the highest number for a progressive party since TEAM's domination in the 70s.

11. Alan Artibise and Michael Seelig, "Planning Research Proved Challenge," *Vancouver Sun*, November 9, 1990, A20.

12. Neal Peirce and Curtis Johnson, "The Peirce Report: A Way to Wed Conservation to Development," *Seattle Times*, October 8, 1989, A4.

13. These three types of "persuasion" are typical of ideological production. See Eagleton, *Ideology*, pp. 51–60; Thompson, *Ideology and Modern Culture*, p. 60; Bourdieu, *Outline of a Theory*; Barrett, *The Politics of Truth*, p. 167.

14. Stanley Kwok, "Expo Site Executive Paints Bright Future," *Vancouver Sun*, November 12, 1990, B3.

15. Quoted in Michael Seelig and Alan Artibise, "Growing to Extremes," *Vancouver Sun*, November 10, 1990, B2.

16. Ibid.

17. They included the following: "Increased densities result in lower costs for housing; increased densities do not mean a city full of highrises; increased densities usually lead to livelier street life, shopping opportunities and cultural activities within walking distance of homes; increased densities can lead to the combined benefits of lower per capita taxes as well as cheaper and better services." In Michael Seelig and Alan Artibise, "Megacity," *Vancouver Sun*, November 12, 1990, B2.

18. Michael Seelig and Alan Artibise, "Land," *Vancouver Sun*, November 14, 1990, B2.

19. "Denmark Likes Closer Quarters," *Vancouver Sun*, November 12, 1990.

20. "Politics," *Vancouver Sun*, November 13, 1990.

21. "The Global Challenge," *Vancouver Sun*, November 15, 1990.

22. Michael Seelig and Alan Artibise, "The Global Challenge," *Vancouver Sun*, November 15, 1990, B2.

23. Hall, "The Toad in the Garden," p. 46; Eagleton, *Ideology*, pp. 14–15.

24. Taylor, *Sources of the Self*.

25. This is part of a broader communitarian critique of liberalism. See, e.g., Lasch, "Communitarian Critique of Liberalism," p. 184.

26. Mouffe, *The Return of the Political*, p. 23.

27. Michael Seelig and Alan Artibise, "Politics," *Vancouver Sun*, November 13, 1990, B2.

28. Ibid.

29. This would follow from the conservative communitarian position of someone like Alasdair MacIntyre, rather than the liberal communitarian tolerance of difference advocated by Charles Taylor. See MacIntyre, *After Virtue*.

30. See, e.g., "Politics," *Vancouver Sun*, November 13, 1990.

31. Quoted in "Growing to Extremes," *Vancouver Sun*, November 10, 1990. In addition to his position as head of the International Finance Center, Goldberg was the dean of the U.B.C. Faculty of Commerce and Business Administration and an economic adviser to Premier Harcourt, Stanley Kwok, Lieutenant-Governor David Lam, and several other politicians and businessmen in British Columbia. In 1985 he wrote *The Chinese Connection*, a book that enumerated increasing Pacific Rim connections and the imperative for Canada to become hooked into powerful Chinese networks of trade and capital. He also wrote numerous articles on global cities, Vancouver real estate, and the Pacific Rim; all emphasized the global nature of the economy and the advantages of integration into the international circuits of capital. See, for example, Goldberg, "Hedging Your Great Grandchildren's Bets" and *The Chinese Connection*.

32. This link is still very much in evidence, as demonstrated by the election of Gordon Campbell as mayor in 1986. Campbell was previously employed by Marathon Realty, the real estate arm of the Canadian Pacific Railroad. See Gutstein, *Vancouver Ltd.*

33. See Macdonald, "CPR Town," p. 385, and "The Canadian Pacific Railway," pp. 422–425.

34. Stam, "Steel of Empire," pp. 33–52; Foran, "CPR and the Urban West," pp. 89–106.

35. Canadian Pacific Officer's Luncheon Club of Montreal, *Canadian Pacific History: Four Addresses*, pamphlet, 1935–1936.

36. By 1912, of the sixteen individuals sitting on the council, at least ten were involved in real estate or business. See Macdonald, "CPR Town," p. 406.

37. Roy, *Vancouver*, p. 42.

38. Of the approximately 300 business leaders in Vancouver at the turn of the century, over 90 percent were from Canada, Britain, or the United States; of these, the vast majority were Protestant, mainly Presbyterian or Anglican. See Macdonald, "CPR Town;" and McDonald, "Working Class Vancouver."

39. Weaver, "The Property Industry," p. 429.

40. For example, Chinese laundromats were often excluded from west-side suburban commercial strips on the grounds of health violations. These same violations did not seem to apply to Chinese laundromats in east-side neighborhoods. See ibid., p. 431; K. Anderson, *Vancouver's Chinatown*.

41. K. Anderson, *Vancouver's Chinatown*.

42. Roy, *Vancouver*, p. 68.

43. Status was defined by social prestige. In 1914, 80 percent of the members of Vancouver's social register lived in Shaughnessy Heights. City of Vancouver, Planning Department, *First Shaughnessy Plan Background Report*, May 1982, p. 3.

44. City of Vancouver, Engineering Department, *Shaughnessy Heights Building Restriction Act, 1922.*

45. The 1922 act was amended in 1927, 1933, and in 1939. Through the lobbying of the Shaughnessy Heights Property Owners Association, the act's restrictions were extended until 1970. The SHPOA also petitioned the government to include a provision whereby SHPOA could register complaints against offenders of the act. See the amendments from 1927, 1933, 1939, and 1950 in ibid. Part of the original rationale behind the formation of the SHPOA was the protection of Shaughnessy's single-family housing, which was threatened by the subletting of large homes as rooming houses during the depression, and by the permission of multifamily dwellings during the war.

46. Bartholomew, A Plan, pp. 89–91; Roy, Vancouver, p. 401.

47. Bartholomew, A Plan, p. 169

48. Barman, "Neighborhood and Community," pp. 114–115.

49. The Municipal Act of 1881 enabled this type of exclusionary development by giving power to municipal councils to regulate the construction of dwelling units and to limit the number of occupants per unit. This power was used frequently in crackdowns on Chinese residences, where Chinese laborers were often forced to live in extremely dense conditions. The Municipal Clauses Act of 1908 increased the power of municipal councils to regulate the location and construction of buildings such as laundromats and washhouses. As discussed earlier, this act was also employed to exclude Chinese people and Chinese businesses from particular neighborhoods. For a list of early land regulations in British Columbia, see S. Hamilton, Regulation.

50. Weaver, "The Property Industry," p. 430.

51. For example, the secretary of the North Vancouver Planning Commission was also the president of North Vancouver Realty. The chair of Victoria's Zoning Committee also worked at Pemberton and Son, Real Estate Co., etc. For more details on the early linking of property ownership and municipal planning, see Weaver, "The Property Industry," p. 433; Gutstein, Vancouver, Ltd.

52. The quote is from Professor Buck. Quoted in Bartholomew, A Plan, p. 300. In July 1911, B.C. Magazine advertised Point Grey as the new "fashionable" area of Vancouver, equating the area with the sophisticated suburban areas of Montreal (Westmount) and Toronto (Rosedale).

53. The Vancouver director of planning wrote of the role of the SHPOA: "Early initiatives undertaken by the Shaughnessy Heights Property Owners' Association were instrumental in drawing City Council's attention to the need for a new plan in Shaughnessy. SHPOA played a key role in responding to those pressures threatening to destroy the historic and aesthetic character of First Shaughnessy by developing a set of goals and recommendations for the areas." City of Vancouver, Planning Department, First Shaughnessy Plan Background Report, p. ii.

54. Later this became the First Shaughnessy District By-law No. 5542. City of Vancouver, Zoning and Development By-laws, First Shaughnessy District, December 1989.

55. The explicit terms of reference of this panel are: "To advise the Development Permit Board or the Director of Planning, as the case may be, regarding all major development permit applications in the First Shaughnessy District; to preserve and protect the heritage and special character of the First Shaughnessy District; (and) to advise the Director of Planning concerning the implementation and effectiveness of the approved planning policies, regulations and design guidelines for the First Shaughnessy District." City of Vancouver, Planning Department, *First Shaughnessy Advisory Design Panel*, March 1988.

56. City of Vancouver, Planning Department, *First Shaughnessy Design Guidelines*, October 1987.

57. Benny Hsieh Chung-pang, the founder of the ad hoc committee, noted sarcastically that the refuse collection memo had been translated into six languages, whereas the SHPOA-designed questionnaire was written only in English.

58. For a discussion of this kind of representational reversal and its power dynamics in black culture, see hooks, "Representing Whiteness."

59. This testimony at the public hearing of October 5, 1992, is cited in Ley, "Between Europe and Asia";for a further discussion of the arguments presented at these hearings see, in particular, pp. 196–199.

60. Allejandro, *Hermeneutics*.

61. Habermas, *The Structural Transformation*.

62. Eley, "Nations, Publics, and Political Cultures"; Benhabib, "Models of Public Space'; Ryan, "Gender and Public Access"; Fraser, *Unruly Practices*; Mansbridge, "Feminism and Democracy."

63. Fraser, *Unruly Practices*, p. 65.

64. Ibid., p. 73.

65. Carole Taylor, "From Saving Trees to a Broad, Mushy Intrusion into Every Single Family Lot?" *Vancouver Sun*, January 21, 1991, A2.

66. Malkki, *"National Geographic,"* p. 27.

67. Urban redevelopment transformed all of Vancouver in the late 1980s, but it was protested most vociferously in the upper-class, mainly white neighborhoods of the city's west side.

68. Quoted in Tom Arnold, "Sequoias Planted in City Park to Mark Loss of Two Giant Trees," *Vancouver Sun*, May 4, 1990, B1.

69. Duncan and Duncan, "A Cultural Analysis."

70. Canetti, *Crowds and Power*, pp. 84–85.

71. Quoted in Mike Usinger, "Huge Trees Threatened," *Vancouver Courier*, March 11, 1990, p. 1.

72. Joanne Blain, "Couple Turned Activists When Dozers Arrived," *Vancouver Sun*, April 16, 1990, B4. For a further analysis of this tree confrontation, see Ley, "Between Europe and Asia." The controversy over the removal of trees on private property has been ongoing and remains one of the most divisive and bitter issues in Vancouver city politics.

73. Malkki, *"National Geographic."*

74. Of course this idea of the country and its ethically superior ways of life was promulgated by suburban residents well within municipal boundaries. The irony of the use of morality in the tree debate extends in several directions, including the fact that a number of the wealthy west-side residents involved in this controversy may have secured their positions in these upscale wooded neighborhoods as a result of profits reaped from the British Columbia timber industry. In addition, as Willems-Braun has shown with regard to the wider rhetoric surrounding the "rainforest" and "nature" in British Columbia, discourses such as these have completely excluded or appropriated the history, experiences, and voices of First Nations people, from whom the "wooded" land was originally seized. See Willems-Braun, "Buried Epistemologies."

75. Here the imperial significance of well-balanced, well-*tended* gardens can be linked with the broader domestication of the landscape under colonialism. The "ideal" landscape thus is one tamed by a "civilizing" force, yet which simultaneously erases the scars of that force through the promotion of equilibrium, stasis, and tradition. See, e.g., Kolodny, *The Lay of the Land*; and Schaffer, *Women and the Bush*.

76. The negative depiction of the city and the linking of the city with a particular group of people was similarly employed against the Jews in fascist Germany, when capitalist practices, Jewish lifestyles, and immoral behavior were linked and located in an urban forum. See Theweleit, *Male Fantasies*.

77. Malkki, "*National Geographic*," p. 34; Deleuze and Guattari, *A Thousand Plateaus*, p. 18.

78. Benedict Anderson writes of this imagined community: "If nation-states are widely conceded to be 'new' and 'historical,' the nations to which they give political expression always loom out of an immemorial past, and, still more important, glide into a limitless future." See B. Anderson, *Imagined Communities*, pp. 11–12.

79. See, e.g., G. Rose, *Feminism and Geography*; and Haraway, *Primate Visions*.

80. Here the feminized feature of landscape production is most evident. The trees that are being protected are coded as female and contain the "seeds" of the future generation. For further discussion of discourses about community and stewardship and the feminization of nature, see G. Rose, *Feminism and Geography*; Merchant, *The Death of Nature*.

81. Quoted in Blain, "Couple Turned Activists." Simmons was a member of the Granville-Woodlands Property Owners Association (GWPOA), one of the main organizations protesting the removal of mature trees on private property. (Note: Because of the sensitive nature of this material, I have used a pseudonym here and in Chapter 2 for the "Simmonses" despite their public role as spokespersons.) Several other urban community groups that protested the changes in Vancouver's landscape during this period were composed of local area residents, primarily property owners. For more information about such organizations, which varied somewhat in style and general

aims, see Majury, "Identity"; Ley, "Between Europe and Asia"; and Pettit, "Zoning."

82. Jack Moore, "Sequoia Saga Makes Waves across Pacific: Hong Kong Media Feast on Local Fallen Tree Story," *Vancouver Courier,* May 2, 1990, p. 8.

83. These letters are discussed in R. A. Rabnett and Associates, "Trees on Single Family Lots: A Program for the Protection of Trees on Private Property," 1990, available from the City of Vancouver Planning Department.

84. Ibid., p. 1.

85. Malkii, *"National Geographic,"* p. 31.

86. Mehta, *Liberalism and Empire.*

## CHAPTER FIVE

1. Gilroy, *Against Race,* p. 98.

2. Rosemary Marangoly George, "Recycling: Long Routes to and from Domestic Fixes," in George, *Burning Down the House,* pp. 1–20.

3. Rybczynski, *Home.*

4. Comaroff and Comaroff, "Home-Made Hegemony," p. 38.

5. Rafael, "Colonial Domesticity." See also Comaroff and Comaroff, "Home-Made Hegemony."

6. See, e.g., Rosemary Marangoly George, "Homes in the Empire, Empires in the Home," in George, *Burning Down the House,* pp. 47–74.

7. T. Mitchell, *Colonising Egypt,* p. 2.

8. Comaroff and Comaroff, "Home-Made Hegemony."

9. For the former, see Hall, "The Toad in the Garden." For an insightful analysis of the latter, see Asad, "Multiculturalism and British Identity."

10. Powell said in Southall in 1971: "It is . . . truly when he looks into the eyes of Asia that the Englishman comes face to face with those who would dispute with him the possession of his native land." Cited in Gilroy, *There Ain't No Black,* p. 45. Hague, as the Tory candidate for prime minister in 2001, opted not to discipline a Tory MP, John Townsend, who gave a highly inflammatory speech about immigration that year. The speech was full of thinly veiled equations of immigrants, people of color, and the "loss" of Britain. Hague's defense of Townsend was seen as a defense of his political views, although from a politically more ambiguous, and thus safer, vantage point.

11. In political terms, it is important to note that the Privy Council in London remained technically the highest level of Canadian government until the Canadian Charter of Rights was legislated in the Canadian Constitution Act of 1982. This legislation symbolized the first full expression of Canadian autonomy from Britain. The legal system, however, remains based on the English court system. Economically, until the 1980s, Britain led as the greatest source of immigrants and foreign direct investment, and there has always been a strong trading relationship between Canada and Britain. Aside from Québec, the culture of Canada is overwhelmingly British, as reflected in language, social and institutional organization, and popular culture. See Wardhaugh, *Language and Nationhood.*

12. Immigration patterns shifted markedly after the 1970s, mainly as a result of the increased immigration from Hong Kong, Taiwan, and other countries in East and Southeast Asia.

13. See, e.g., Ward, *White Canada Forever*.

14. See Harris, *The Resettlement of British Columbia*, esp. chapter 1.

15. The legislative assembly was established on Vancouver Island in 1849, and the B.C. provincial parliament was formed in 1871. See K. Anderson, *Vancouver's Chinatown*, p. 24.

16. Ibid., p. 26.

17. Ibid., p. 29.

18. The internment of Japanese Canadians during World War I was a more egregious but shorter-term example of spatial confinement based on racial definition. Immigration from China was greatly curtailed in 1885 with the first Chinese immigration act (the Act to Restrict and Regulate Chinese Immigration), which imposed a fifty-dollar head tax and limited the number of immigrants per ship. The head tax increased in successive years. In 1923, all immigration from China was halted with the Chinese Exclusion Act, which was not repealed until 1947. In 1962 the country of origin was removed as a major criterion for admission, but it was not until 1967 that immigration policy was completely overhauled and the last "officially" discriminatory elements were removed.

19. K. Anderson, *Vancouver's Chinatown*.

20. One realtor told me that restrictive covenants based on race were operating in North Vancouver well into the 1980s. Author's interview, Vancouver, November 1990.

21. This relationship was also impacted by the differential relations of power between Britain and China during this time.

22. The maximum height and square footage depended on the size of the lot and the date of house construction. The height limit was thirty-five feet before April 1986, and thirty feet after April 1988.

23. The standard pattern in Vancouver single-family neighborhoods is to have the garage in back, with a separate alley for access.

24. Interviews with Annie Humphreys took place between October and December 1990. Bracketed sections refer to street locations in Vancouver. (Note: Because of the sensitive nature of this material, I have used a pseudonym here and in Chapter 2 for "Annie" despite her position as an official spokesperson for CCAH.)

25. This group was concerned with monster houses and overall neighborhood change but was especially active in protesting the demolition of low-rise apartment buildings in Kerrisdale.

26. Fong discussed similar struggles over the linking of racism and slow-growth development movements in his study of the transformation of Monterey, California, in the early 1990s. See Fong, *The First Suburban Chinatown*.

27. George, "Recycling." p. 3.

28. Niall Majury photographed this in 1990. See Majury, "Identity."

29. The phrase may also come from book 5 of Virgil's *Aeneid,* where he speaks of the "genius of the place" in reference to the individual and pre-existing spirituality of particular places. This idea of preexisting spiritual features was used by eighteenth-century English landscape designers, who attempted to build from and on the original and existing features of a site rather than remold the landscape completely. This English landscape tradition is the basis for most of the garden designs in west-side Vancouver.

30. See, e.g., McClintock, *Imperial Leather;* Ware, *Beyond the Pale;* and Burton, *Dwelling in the Archive* and *At the Heart.*

31. George, "Homes in the Empire," p. 48.

32. Hansen, "Introduction," pp. 16–17.

33. Spivak uses this phrase in reference to Foucault's studies of the spaces of prisons (neglecting the wider socioeconomic ramifications of colonialism). Spivak, "Can the Subaltern Speak?" p. 291. See also Gilroy, *Against Race,* p. 65.

34. Silverman, *The Subject of Semiotics.*

35. Bristol discusses how architectural style was used as the scapegoat for the numerous problems associated with the Pruitt-Igoe public housing project in St. Louis. By focusing on "bad design," problems of chronic racism, poverty, unequal access to resources, and general social and economic breakdown were displaced and ignored. See Bristol, "The Pruitt-Igoe Myth."

36. Gilroy noted for the British case that the fear of competition over limited resources was exacerbated by the depiction of black people as abnormally fecund; i.e., by "outproducing" white people they would be able to attain greater access to desirable resources. See Gilroy, *Ain't No Black in the Union Jack.*

37. The term "enablers" is from Gutstein, *The New Landlords.*

38. Kristeva, *Strangers to Ourselves,* p. 103.

39. Holdsworth, "Cottages and Castles," p. 28.

40. Ibid.

41. City of Vancouver, Planning Department, *Vancouver Local Areas, 1986* (1988). The sample is 100 percent data from the Canada Census of 1986. For Shaughnessy statistics, see also "Vancouver's Housing: Housing Stock," p. 11. As census areas do not correspond exactly to local areas in this publication, figures are approximate.

42. The latter two styles derived from the Arts and Crafts Movement highly popular in Britain and North America in the 1920s. The bungalows are typically one-story wooden buildings with a broad verandah, shingled siding, and an informal indoor-outdoor plan. The cottages are stucco, with some half-timber trim. For detailed descriptions of these house styles and their architectural history in Vancouver, see Holdsworth, "Cottages and Castles," pp. 29–30.

43. Duncan and Duncan, "A Cultural Analysis," p. 270.

44. Craig Spence, quoted in "Being Neighbourly—By Law," *Vancouver Courier,* December 18, 1985, p. 7.

45. Comaroff and Comaroff, "Home-Made Hegemony," p. 39.

46. Turner, *Capability Brown*, p. 33.

47. See Girouard, *English Country House.*

48. For a similar analysis related to eighteenth-century painting, see Berger, *Ways of Seeing.*

49. Duncan and Duncan, "A Cultural Analysis," p. 260; Weiner, *English Culture.*

50. Compusearch Market and Social Research Ltd., "Assets and Indebtedness: Dollar Values," 41st and West Boulevard (Vancouver) (Benchmark: Vancouver Census Metropolitan Area), October 29, 1987.

51. The letter was one of forty-two letters and briefs the Planning Department received in February and March commenting on the RS-1 proposal; 90 percent supported the general intent to limit house size. See City of Vancouver, Planning Department, *1988: RS-1 Proposals: Public Consultation and Submissions*, 1988.

52. Hiebert, "Immigration," p. 69.

53. Bourdieu, *In Other Words.*

54. Bourdieu argues that legitimating culture as second nature allows those with it to see themselves as disinterested and unblemished by any mercenary uses of culture. See Bourdieu, *Distinction*, p. 86.

55. Ibid., p. 124.

56. Author's interview with a Hong Kong Chinese woman planning to emigrate to Vancouver, Hong Kong, May 1991.

57. Hannerz, "Cosmopolitans and Locals," p. 237.

58. The half-hour documentary, entitled "The Hong Kong Connection," was produced by Radio-Television Hong Kong. In the film, white-first sentiments expressed by columnist Doug Collins manifested a particularly harsh strain of racism in Vancouver.

59. See, e.g., Collins, "Vancouver Feels the Asian Strain" "Chuppies Become Latest Target of Vancouver's Racism," *Hong Kong Standard*, May 20, 1989; "Welcome Wears Thin for 'The Yacht People,'" *South China Morning Post* (Hong Kong), January 13, 1990.

60. Hannerz, "Cosmopolitans and Locals."

61. Williams, *Towards 2000*, p. 195.

62. Gilroy, *There Ain't No Black*, pp. 49–50.

63. Lowe, *Immigrant Acts*; Ong, "Cultural Citizenship as Subject-Making."

64. See, e.g., Lau, *Society and Politics*, p. 69.

65. Sally is the daughter of a wealthy Shanghai businessman who emigrated to Hong Kong in the late 1940s. She is married to a prominent physician, the son of another wealthy Hong Kong family.

66. Bourdieu, *Distinction*, p. 57.

67. Knapp, *The Chinese House*, p. 54.

68. Rossbach, *Interior Design with Feng-Shui*, p. 51.

69. Real-estate agents in Hong Kong confirmed this perception and noted that as a general rule, their clients who had average incomes (middle to

upper middle class in Vancouver terms) tended to buy in Richmond, where new houses were large but prices were slightly lower than in the west-side neighborhoods.

70. Calvino, *Invisible Cities*, p. 10.

71. George, "Recycling."

72. See, e.g., the special series of *The Economist*, September 9–15, 1995, on the decline of the family. This series, discussed in detail by George in "Recycling," pp. 6–7, pictures the father at the center in a "roseate yet clearly patriarchal view of past domestic arrangements."

73. A few examples: *China Tide*, a popular book by Margaret Cannon in 1989, chronicling Hong Kong investment in Vancouver; "Asian Capital: The Next Wave," *B.C. Business*, July 1990; "Tidal Wave from Hong Kong," *B.C. Business*, February 1989; "Flippers Awash in Profits"; "Hong Kong Capital Flows Here."

74. Theweleit, *Male Fantasies*.

75. Gordon Hamilton and Daphne Bramham, "Death of the Middle Class," *Vancouver Sun*, November 14, 1992, p. A1.

76. The phrase is from Harrison and Bluestone, *The Great U-Turn*.

77. I am not arguing that these spaces literally "protected" men or women, but rather that this *ideology* of coherent and protective domestic spaces was widespread, and was disrupted by the various neoliberal practices and their corresponding effects discussed here.

78. Brown, *Domestic Individualism*.

79. Ong, *Flexible Citizenship*, p. 113.

80. Ibid, esp. chapter 4. See also Yeoh and Willis, "'Heart' and 'Wing.'"

81. Ong and Nonini use the term "ungrounded" in the title of their edited volume, *Ungrounded Empires*. The "uncanny" in architecture was often linked with vacant or abandoned buildings. Many of the monster houses remained empty in Vancouver, as the offshore Hong Kong buyers purchased them in advance of their arrival, sometimes by several years. "Linked by Freud to the death drive, to fear of castration, to the impossible desire to return to the womb, the uncanny has been interpreted as a dominant constituent of modern nostalgia, with a corresponding spatiality that touches all aspects of social life." Vidler, *The Architectural Uncanny*, p. x.

## CHAPTER SIX

1. Mehta, *Liberalism and Empire*, p. 118.

2. Ibid., p. 121.

3. See also Holston and Appadurai, "Introduction," pp. 1–12.

4. Mehta, *Liberalism and Empire*; Fraser, *Unruly Practices*; Macpherson, *Political Theory of Possessive Individualism*.

5. Ong and others have labeled these kinds of informal constructions of belonging "cultural citizenship." See Ong, "Cultural Citizenship as Subject-Making" and "Making the Biopolitical Subject."

6. Isin, *Becoming Political*, p. 4.

7. See, e.g., Lisa Lowe, *Immigrant Acts*; Ignatiev, *How the Irish Became White*; Jacobson, *Whiteness of a Different Color*; K. Anderson, *Vancouver's Chinatown*; and Roediger, *The Wages of Whiteness*.

8. For just a few examples of what is now a very large literature, see Castles and Davidson, *Citizenship and Migration*; Cheah and Robbins, *Cosmopolitics*; Glick Schiller and Fouron, *Georges Woke Up Laughing*; Münch, *Nation and Citizenship*.

9. See my discussion of the transformation of multiculturalism in education: K. Mitchell, "Educating the National Citizen."

10. Bourdieu and Wacquant, "NewLiberalSpeak," p. x.

11. Mehta, *Liberalism and Empire*; Pateman, *The Disorder of Women*; Lowe, *Immigrant Acts*; Isin, *Becoming Political*.

12. Gilroy, *Against Race*, p. 65.

13. Coronil, "Towards a Critique of Globalcentrism."

# Bibliography

Abu-Lughod, Janet. *Before European Hegemony: The World System A.D. 1250–1350.* Oxford: Oxford University Press, 1989.

Agnew, John, and Stuart Corbridge. *Mastering Space: Hegemony, Territory, and International Political Economy.* New York: Routledge, 1995.

Alibhai-Brown, Yasmin. *After Multiculturalism.* London: Foreign Policy Center, 2001.

Allejandro, Robert. *Hermeneutics, Citizenship, and the Public Sphere.* Albany: State University of New York Press, 1993.

Althusser, Louis. "Ideology and Ideological State Apparatuses: Notes towards an Investigation." *Lenin and Philosophy and other Essays.* Translated by Ben Brewster. New York: Monthly Review Press, 1971.

Anderson, Benedict. *Imagined Communities: Reflections on the Origin and Spread of Nationalism.* London: Verso, 1991.

Anderson, Kay. *Vancouver's Chinatown: Racial Discourse in Canada, 1875–1980.* Montreal: McGill-Queen's University Press, 1991.

Anderson, Perry. "The Antinomies of Antonio Gramsci." *New Left Review* 100, pp. 5–78.

Anderson, Robert, and Eleanor Wachtel, eds. *The Expo Story.* Madeira Park, B.C.: Harbour, 1986.

Appadurai, Arjun. *Modernity at Large: Cultural Dimensions of Globalization.* Minneapolis: University of Minnesota Press, 1996.

Asad, Talal. "Multiculturalism and British Identity in the Wake of the Rushdie Affair." *Politics and Society* 18, 4 (1990): 455–480.

Barman, Jean. "Neighborhood and Community in Interwar Vancouver: Residential Differentiation and Civic Voting Behavior." In *Vancouver Past: Essays in Social History,* ed. Robert McDonald and Jean Barman, pp. 97–141. Vancouver: University of British Columbia Press, 1986.

Barrett, Michèle. *The Politics of Truth: From Marx to Foucault.* Stanford: Stanford University Press, 1991.

Bartholomew, Howard. *A Plan for the City of Vancouver.* Vancouver: City of Vancouver, 1928.

Baumann, Gerd. *The Multicultural Riddle: Rethinking National, Ethnic, and Religious Identities.* New York: Routledge, 1999.

Beaverstock, Jonathan. "Transnational Elites in Global Cities: British Expatriates in Singapore's Financial District." *Geoforum* 33 (2002): 525–538.

Beck, Ulrich. *What Is Globalization?* Translated by Patrick Camiller. Cambridge: Polity Press, 2000.

Behiels, Michael. *Prelude to Quebec's Quiet Revolution: Liberalism versus Neo-Nationalism, 1945–1960.* Montreal: McGill-Queen's University Press, 1985.

Benhabib, Seyla. *The Claims of Culture: Equality and Diversity in the Global Era.* Princeton: Princeton University Press, 2002.

———. "Models of Public Space: Hannah Arendt, the Liberal Tradition, and Jurgen Habermas." In *Situating the Self,* ed. Seyla Benhabib, pp. 89–120. New York: Routledge, 1992.

Bennett, David. "Introduction." In *Multicultural States: Rethinking Difference and Identity,* ed. David Bennett, pp. 1–25. London: Routledge, 1998.

Berger, John. *Ways of Seeing.* London: British Broadcasting Corporation and Penguin Books, 1972.

Berman, Marshall. *All That Is Solid Melts into Air: The Experience of Modernity.* New York: Viking Penguin, 1988.

Bhabha, Homi. "Culture's In Between." In *Multicultural States: Rethinking Difference and Identity,* ed. David Bennett, pp. 29–36. London: Routledge, 1998.

———. *Nation and Narration.* New York: Routledge, 1990.

Bouraoui, Hédi. *The Canadian Alternative: Cultural Pluralism and the Canadian Unity.* Downsview: ECW Press, 1979.

Bourdieu, Pierre. *Distinction: A Social Critique of the Judgement of Taste.* Translated by Richard Nice. Cambridge: Harvard University Press, 1984.

———. *In Other Words: Essays Towards a Reflexive Sociology.* Translated by Matthew Adamson. Palo Alto: Stanford University Press, 1990.

———. *Outline of a Theory of Practice.* Cambridge: Harvard University Press, 1977.

Bourdieu, Pierre, and Louis Wacquant. "NewLiberalSpeak: Notes on the New Planetary Vulgate." *Radical Philosophy* 105 (2001): 2–5.

Brennan, Teresa. *Globalization and Its Terrors: Daily Life in the West.* London: Routledge, 2003.

Brenner, Neil. "State Territorial Restructuring and the Production of Spatial Scale: Urban and Regional Planning in the Federal Republic of Germany, 1960–1990," *Political Geography* 16, 4 (1997): 273–306.

Brenner, Neil, and Neil Theodore. "Cities and Geographies of 'Actually Existing Neoliberalism.'" *Antipode* 34, 3 (2002): 349–379.

Bristol, Katharine. "The Pruitt-Igoe Myth." *Journal of Architectural Education* 44, 3 (1991): 163–171.

Brown, Gillian. *Domestic Individualism: Imagining Self in 19th Century America.* Berkeley: University of California Press, 1990.

Bullivant, Brian. *Pluralism: Cultural Maintenance and Evolution.* Clevedon, England: Multilingual Matters, 1984.

Burton, Antoinette. *At the Heart of the Empire: Indians and the Colonial Encounter in Late-Victorian Britain.* Berkeley: University of California Press, 1998.

———. *Dwelling in the Archive: Women Writing House, Home, and History in Late Colonial India.* Oxford: Oxford University Press, 2003.

Cairns, Alan. *Constitution, Government, and Society in Canada.* Toronto: McClelland and Steward, 1988.

Calvino, Italo. *Invisible Cities.* Translated by William Weaver. New York: Harcourt Brace Jovanovich, 1978.

Canada Mortgage and Housing Corporation. "Vacancy Rates in Privately Initiated Rental Apartment Structures of Six Units and Over, by Number of Bedrooms, 1989–1990." *Canadian Housing Statistics 1990.* Ottawa: CMHC.

*The Canadian Multiculturalism Act: A Guide for Canadians.* Ottawa: Minister of Supply and Services Canada, 1990.

Canetti, Elias. *Crowds and Power.* Translated by Carol Stewart. New York: Seabury Press, 1978.

Cannon, Margaret. *China Tide: The Revealing Story of the Hong Kong Exodus to Canada.* Toronto: Harper and Collins, 1989.

Castells, Manuel. *The Rise of the Network Society.* Oxford: Blackwell, 1996.

Castles, Stephen, and Alastair Davidson. *Citizenship and Migration: Globalization and the Politics of Belonging.* London: Macmillan Press, 2000.

Chakrabarty, Dipesh. "Modernity and Ethnicity in India." In *Multicultural States: Rethinking Difference and Identity,* ed. David Bennett, pp. 91–110. London: Routledge, 1998.

Cheah, Pheng, and Bruce Robbins. *Cosmopolitics: Thinking and Feeling Beyond the Nation.* Minneapolis: University of Minnesota Press, 1998.

City of Vancouver. "Apartment Vacancy Rates, 1976–1992." *Vancouver Monitoring Program.* City of Vancouver, August 1992.

Clarkson, Steven. "Constitutionalizing the Canadian-American Relationship." In *Canada under Free Trade,* ed. Duncan Cameron and Mel Watkins, pp. 3–20. Toronto: James Lorimer, 1993.

Clement, Wallace. *The Canadian Corporate Elite.* Ottawa: McClelland and Stewart, 1975.

Comaroff, John, and Jean Comaroff. "Home-made Hegemony: Modernity, Domesticity, and Colonialism in South Africa." In *African Encounters with Domesticity,* ed. Karen Hansen, pp. 37–74. New Brunswick: Rutgers University Press, 1992.

Connolly, William. *The Ethos of Pluralization.* Minneapolis: University of Minnesota Press, 1995.

Coronil, Fernando. "Towards a Critique of Globalcentrism: Speculations on Capitalism's Nature," *Public Culture* 12, 2 (2000): 351–374.

Cox, Kevin. *Spaces of Globalization: Reasserting the Power of the Local.* New York: Guilford, 1997.

Creighton, Donald. *Canada's First Century.* Toronto: Macmillan, 1970.

Cruz, Jon. "From Farce to Tragedy: Reflections on the Reification of Race at Century's End." In *Mapping Multiculturalism,* ed. Avery Gordon and Christopher Newfield, pp. 19–39. Minneapolis: University of Minnesota Press, 1996.

Cybriwsky, Roman, David Ley, and John Western. "The Political and Social Construction of Revitalized Neighborhoods: Society Hill, Philadelphia, and False Creek, Vancouver." In *Gentrification of the City,* ed. Neil Smith and Peter Williams, pp. 92–120. Boston: Unwin Hyman, 1986.

Dalton, Greg. "Canada: Who Benefits from Investor Immigration?" *Hong Kong Business*, July 1991.

Davis, Mike. *City of Quartz: Excavating the Future in Los Angeles*. New York: Verso, 1990.

Deleuze, Gilles, and Felix Guattari. *A Thousand Plateaus: Capitalism and Schizophrenia*. Translated by Brian Massumi. Minneapolis: University of Minnesota Press, 1987.

Dewey, John. *John Dewey, The Later Works, 1925–1953*. Edited by Jo Ann Boydston. Carbondale: Southern Illinois University Press, 1987.

Dicken, Peter. *Global Shift: The Internationalization of Economic Activity*. New York: Guilford Press, 1992.

Dirlik, Arif. "Chinese History and the Question of Orientalism." *History and Theory* 35, 4 (1996): 96–118.

Donzelot, Jacques. "Pleasure in Work." In *The Foucault Effect: Studies in Governmentality*, ed. Gordon Burchell, Colin Gordon, and Peter Miller, pp. 251–280. Chicago: University of Chicago Press, 1991.

Drache, Daniel. "The Future of Trading Blocs." In *Canada under Free Trade*, ed. Duncan Cameron and Mel Watkins, pp. 264–276. Toronto: James Lorimer, 1993.

Duncan, James, and Nancy Duncan. "A Cultural Analysis of Urban Residential Landscapes in North America: The Case of the Anglophile Elite." In *The City in Cultural Context*, ed. John Agnew, John Mercer, and David Sopher, pp. 255–276. Boston: Allen and Unwin, 1984.

Eagleton, Terry. *Ideology*. London: Verso, 1991.

Eddy, John, and Deryck Schreuder. "Canadians, Canadiens, and Colonial Nationalism, 1896–1914: The Thorn in the Lion's Paw." In *The Rise of Colonial Nationalism: Australia, New Zealand, Canada, and South Africa First Assert Their Nationalities, 1880–1914*, ed. John Eddy and Deryck Schreuder, pp. 157–183. Sydney: Allen and Unwin, 1988.

Eley, Geoff. "Nations, Publics, and Political Cultures: Placing Habermas in the Nineteenth Century." In *Habermas and the Public Sphere*, ed. Craig Calhoun, pp. 289–339. Cambridge: MIT Press, 1992.

Elliot, Jean, and Augie Fleras. "Immigration and the Canadian Ethnic Mosaic." In *Race and Ethnic Relations in Canada*, ed. Peter Li, pp. 51–76. Toronto: Oxford University Press, 1990.

Elshtain, Jean. *Public Man, Private Woman: Women in Social and Political Thought*. Princeton: Princeton University Press, 1981.

Employment and Immigration Canada. *Annual Immigration Statistics*. Ottawa: Minister of Supply and Services Canada, 1999.

———. *Doing Business in Canada: A Guide to Canada's Business Immigration Program*. Ottawa: Minister of Supply and Services Canada, 1989.

———. *Immigration Regulations, Guidelines, and Procedures: Business Immigration Program*. Ottawa: Minister of Supply and Services Canada, 1989.

———. *Immigration to Canada: A Statistical Overview*. Ottawa: Minister of Supply and Services Canada, 1989.

Fei, Xiaotong. *From the Soil: The Foundations of Chinese Society.* Translated by Gary Hamilton and Wei Zheng. Berkeley: University of California Press, 1992.

Fong, Timothy. *The First Suburban Chinatown: The Remaking of Monterey Park, California.* Philadelphia: Temple University Press, 1994.

Foran, Max. "The CPR and the Urban West, 1881–1930." In *The CPR West: The Iron Road and the Making of a Nation,* ed. Hugh Dempsey, pp. 89–106. Vancouver: Douglas and McIntyre, 1984.

Frankel, Boris. "Confronting Neo-Liberal Regimes: The Post-Marxist Embrace of Populism and Realpolitik." *New Left Review* 226 (1997): 7–92.

Fraser, Nancy. "Rethinking the Public Sphere: A Contribution to the Critique of Actually Existing Democracy." *Social Text* 25/26 (1990): 56–80.

———. *Unruly Practices: Power, Discourse, and Gender in Contemporary Social Theory.* Minneapolis: University of Minnesota Press, 1989.

Fukuyama, Francis. *The End of History and the Last Man.* New York: Free Press, 1992.

George, Rosemary Marangoly, ed. *Burning Down the House: Recycling Domesticity* Boulder: Westview Press, 1998.

Gereffi, Gary, and Miguel Korzeniewicz. *Commodity Chains and Global Capitalism.* Westport: Praeger, 1994.

Gilroy, Paul. *Against Race: Imagining Political Culture Beyond the Color Line.* Cambridge: Harvard University Press, 2000.

———. *There Ain't No Black in the Union Jack: The Cultural Politics of Race and Nation.* Chicago: University of Chicago Press, 1991.

Girouard, Mark. *Life in the English Country House.* New York: Penguin Books, 1980.

Glick Schiller, Nina, and Georges Fouron. *Georges Woke Up Laughing: Long Distance Nationalism and the Search for Home.* Durham: Duke University Press, 2001.

Glick Schiller, Nina, Linda Basch, and Cristina Szanton Blanc. *Nations Unbound: Transnational Projects, Postcolonial Predicaments, and Deterritorialized Nation-States.* New York: Gordon and Breach, 1994.

———. *Towards a Transnational Perspective on Migration: Race, Class, Ethnicity, and Nationalism Reconsidered.* New York: New York Academy of Sciences, 1992.

Goldberg, Michael. *The Chinese Connection: Getting Plugged In to Pacific Rim Real Estate, Trade, and Capital Markets.* Vancouver: University of British Columbia Press, 1985.

———. "Hedging Your Great Grandchildren's Bets: The Case of Overseas Chinese Investment in Real Estate around the Cities of the Pacific Rim." In *Canada and the Changing Economy of the Pacific Basin.* (Working Paper No. 22.) Vancouver: University of British Columbia Institute of Asian Research, 1984.

Gordon, Avery, and Chrisopher Newfield. Introduction to *Mapping Multiculturalism,* ed. Avery Gordon and Christopher Newfield, pp. 1–17. Minneapolis: University of Minnesota Press, 1996.

Gramsci, Antonio. *Selections from the Prison Notebooks.* Translated by Quintin Hoare and Geoffrey Smith. New York: International, 1997 [1971].

Gray, John. *Liberalism.* Milton Keynes, England: Open University Press, 1986.

Guarnizo, Luis. "The Rise of Transnational Social Formations: Mexican and Dominican State Responses to Transnational Migration." *Political Power and Social Theory* 12 (1998): 45–94.

Guarnizo, Luis, and Michael Peter Smith. "The Locations of Transnationalism." In *Transnationalism from Below,* ed. Michael Peter Smith and Luis Guarnizo, pp. 3–34. New Brunswick/London: Transaction, 1999.

Gupta, Akhil, and James Ferguson. "Beyond 'Culture': Space, Identity, and the Politics of Difference." *Cultural Anthropology* 7, 1 (1992): 6–23.

Gutstein, Donald. "Expo's Impact on the City." In *The Expo Story,* ed. Robert Anderson and Eleanor Wachtel, pp. 65–99. Madeira Park, B.C.: Harbour, 1986.

————. *The New Landlords: Asian Investment in Canadian Real Estate.* Victoria, B.C.: Porcepic Books, 1990.

————. *Vancouver, Ltd.* Toronto: J. Lorimer, 1975.

Habermas, Jurgen. *The Structural Transformation of the Public Sphere: An Inquiry into a Category of Bourgeois Society.* Translated by Thomas Burger and Frederick Lawrence. Cambridge: MIT Press, 1989.

Hall, Stuart. "The Problem of Ideology: Marxism without Guarantees." In *Stuart Hall: Critical Dialogues in Cultural Studies,* ed. David Morley and Kuan-hsing Chen, pp. 25–46. London: Routledge, 1996.

————. "The Toad in the Garden: Thatcherism among the Theorists." In *Marxism and the Interpretation of Culture,* ed. Cary Nelson and Lawrence Grossberg, pp. 35–73. Chicago: University of Chicago Press, 1988.

Hamilton, Gary. Introduction to *From the Soil: The Foundations of Chinese Society,* ed. Gary Hamilton. Berkeley: University of California Press, 1992.

Hamilton, Gary, and Edward Chen. "Introduction: Business Groups and Economic Development." In *Asian Business Networks,* ed. Gary Hamilton, pp. 1–6. Berlin: Walter de Gruyter, 1996.

Hamilton, S. W. *Regulation and Other Forms of Government Intervention Regarding Real Property.* Technical Report No. 13, pp. 67–69. Economic Council of Canada, University of British Columbia, July 1981.

Hannerz, Ulf. "Cosmopolitans and Locals in World Culture." *Theory, Culture, and Society* 7, 2–3 (1990): 237–251.

Hansen, Karen. "Introduction: Domesticity in Africa." In *African Encounters with Domesticity,* ed. Karen Hansen, pp. 1–33. New Brunswick: Rutgers University Press, 1992.

Haraway, Donna. *Primate Visions.* New York: Routledge, 1989.

Harris, Cole. *The Resettlement of British Columbia: Essays on Colonialism and Geographical Change.* Vancouver: University of British Columbia Press, 1997.

Harrison, Bennett, and Barry Bluestone. *The Great U-Turn: Corporate Restructuring and the Polarizing of America.* New York: Basic Books, 1988.

Harvey, David. "Flexible Accumulation through Urbanization: Reflections on 'Post-Modernism' in the American City." *Antipode* 19, 3 (1987): 260–286.

———. "From Managerialism to Entrepreneurialism: The Transformation in Urban Governance in Late Capitalism." *Geografiska Annaler* 71B, 1 (1989): 3–17.

Held, David, Anthony McGrew, David Goldblatt, and Jonathan Perraton. *Global Transformations: Politics, Economics, and Culture.* Stanford: Stanford University Press, 1999.

Hiebert, Dan. "Immigration and the Changing Social Geography of Greater Vancouver. *BC Studies* 121 (1999): 69–82.

Hirst, Paul, and Grahame Thompson. *Globalization in Question: The International Economy and the Possibilities of Governance.* Cambridge: Polity Press, 1999.

Hobhouse, Leonard. *Liberalism.* Oxford: Oxford University Press, 1964.

Holdsworth, Deryck. "Cottages and Castles for Vancouver Home-seekers." *BC Studies* 69–70 (1986): 11–32.

Holston, James, and Arjun Appadurai. "Introduction: Cities and Citizenship." In *Cities and Citizenship,* ed. James Holston, pp. 1–12. Durham: Duke University Press, 1999.

"Hong Kong Capital Flows Here Ever Faster." *Vancouver Sun,* March 21, 1989.

hooks, bell. "Representing Whiteness in the Black Imagination." In *Cultural Studies,* ed. Lawrence Grossberg, Cary Nelson, and Paula Treichler, pp. 338–346. New York: Routledge, 1992.

Ignatiev, Noel. *How the Irish Became White.* New York: Routledge, 1995.

Isin, Engin. *Becoming Political: Genealogies of Citizenship.* Minneapolis: University of Minnesota Press, 2002.

Jacobson, Matthew. *Whiteness of a Different Color: European Immigrants and the Alchemy of Race.* Cambridge: Harvard University Press, 1998.

Jenson, Jane. "'Different' but not 'Exceptional': Canada's Permeable Fordism." *Canadian Review of Sociology and Anthropology* 26, 1 (1989): 69–93.

Jessop, Bob. "Liberalism, Neoliberalism, and Urban Governance: A State-Theoretical Perspective." *Antipode* 34, 3 (2002): 452–472.

———. "A Neo-Gramscian Approach to the Regulation of Urban Regimes: Accumulation Strategies, Hegemonic Projects, and Governance." In *Reconstructing Urban Regime Theory: Regulating Urban Politics in a Global Economy,* ed. Mickey Lauria, pp. 51–73. Thousand Oaks, Calif.: Sage Publications, 1997.

Kallen, Evelyn. "Multiculturalism: Ideology, Policy, and Reality." *Journal of Canadian Studies* 17, 1 (1982): 51–63.

Katz, Cindi. "Hiding the Target: Social Reproduction in the Privatized Urban Environment." In *Postmodern Geography: Theory and Praxis*, ed. Claudio Minca, pp. 93–110. Oxford: Blackwell, 2001.

———. "Vagabond Capitalism and the Necessity of Social Reproduction." *Antipode* 33, 4 (2001): 709–728.

Katznelson, Ira. *Liberalism's Crooked Circle*. Princeton: Princeton University Press, 1996.

Kearney, Michael. "Borders and Boundaries of State and Self at the End of Empire." *Journal of Historical Sociology* 4 (1991): 52–73.

Keast, Gordon. "The World's Longest Commute." *Equity*, March 1988.

Keil, Roger. "Common-Sense Neoliberalism: Progressive Conservative Urbanism in Toronto, Canada." *Antipode* 34, 3 (2002): 579.

Knapp, Ronald. *The Chinese House: Craft, Symbol, and the Folk Tradition*. Hong Kong: Oxford University Press, 1990.

Knox, Paul. "Capital, Material Culture, and Socio-Spatial Differentiation." In *The Restless Urban Landscape*, ed. Paul Knox, pp. 1–34. Englewood Cliffs, N.J.: Prentice Hall, 1993.

Kobayashi, Audrey. "Multiculturalism: Representing a Canadian Institution." In *Place/Culture/Representation*, ed. David Ley and James Duncan, pp. 205–231. Routledge: London, 1993.

Kolodny, Annette. *The Lay of the Land: Metaphor As Experience and History in American Life and Letters*. Chapel Hill: University of North Carolina Press, 1975.

Kristeva, Julia. *Strangers to Ourselves*. New York: Columbia University Press, 1991.

Kymlicka, Will. *Finding Our Way: Rethinking Ethnocultural Relations in Canada*. Toronto: Oxford University Press, 1999.

———. *Multicultural Citizenship: A Liberal Theory of Minority Rights*. Oxford: Clarendon Press, 1995.

Larner, Wendy. "Neo-Liberalism: Policy, Ideology, Governmentality." *Studies in Political Economy* 63 (2000): 5–25.

Lary, Diana. "Regional Variations in Hong Kong Immigration." *Canada and Hong Kong Update*, Fall 1991.

———. "Trends in Immigration from Hong Kong." *Canada and Hong Kong Update*, Summer 1992.

Lasch, Christopher. "The Communitarian Critique of Liberalism." In *Community in America: The Challenge of Habits of the Heart*, ed. C. H. Reynolds and R. V. Norman, pp. 173–184. Berkeley: University of California Press, 1988.

Lau, Siu-Kai. *Society and Politics in Hong Kong*. Hong Kong: Chinese University Press, 1982.

Ley, David. "Between Europe and Asia: The Case of the Missing Sequoias." *Ecumene* 2, 2 (1995): 185–210.

———. "Liberal Ideology and the Postindustrial City." *Annals of the Association of American Geographers*, 70, 2 (1980): 238–258.

————. "Seeking 'Homo Economicus' Transnationally; The Strange Story of Canada's Business Immigration Programme." RIIM Working Papers, no. 00-02.Vancouver: University of British Columbia, 2000.

————. "Styles of the Times. Liberal and Neo-Conservative Landscapes in Inner Vancouver, 1968–1986." *Journal of Historical Geography* 13 (1987): 40–56.

Li, Peter. "Race and Ethnicity." In *Race and Ethnic Relations in Canada*, ed. Peter Li, pp. 3–17. Toronto: Oxford University Press, 1990.

Lipsett, Seymour. *Continental Divide: The Values and Institutions of the United States and Canada.* New York: Routledge, 1990.

Locke, John. *Two Treatises of Government.* Second edition. Edited by P. Laslett. Cambridge: Cambridge University Press, 1967.

Logan, John. "Cycles and Trends in the Globalization of Real Estate." In *The Restless Urban Landscape*, ed. Paul Knox, pp. 33–54. Englewood Cliffs, N.J.: Prentice Hall, 1993.

Lowe, Lisa. *Immigrant Acts: On Asian American Cultural Politics.* Durham: Duke University Press, 1996.

Lukes, Steven. *Moral Conflict and Politics.* Oxford: Clarendon Press, 1991.

Macdonald, Norbert. "The Canadian Pacific Railway and Vancouver's Development to 1900." In *British Columbia: Historical Readings*, ed. W. Peter Ward and Robert McDonald, pp. 422–425. Vancouver: Douglas and McIntyre, 1981.

————. "CPR Town: The City-Building Process in Vancouver, 1860–1914." In *Shaping the Urban Landscape*, ed. Gilbert Stelter and Alan Artibise, pp. 382–412. Ottawa: Carleton University Press, 1982.

MacIntyre, Alasdair. *After Virtue: A Study in Moral Virtue.* London: Duckworth, 1981.

MacLeod, Gordon. "From Urban Entrepreneurialism to a 'Revanchist City'? On the Spatial Injustices of Glasgow's Renaissance." *Antipode* 34, 3 (2002): 602–624.

Macpherson, C. B. *The Political Theory of Possessive Individualism: Hobbes to Locke.* Oxford: Oxford University Press, 1962.

Majury, Niall. "Identity, Place, Power, and the 'Text': Kerry's Dale and the 'Monster' House." Master's thesis, University of British Columbia, 1990.

Malkki, Liisa. "*National Geographic*: The Rooting of Peoples and the Territorialization of National Identity among Scholars and Refugees." *Cultural Anthropology* 7 (1992): 24–44.

Mandel, Michael. *The Charter of Rights and the Legalization of Politics in Canada.* Toronto: Wall and Thompson, 1989.

Mansbridge, Jane. "Feminism and Democracy." *American Prospect*, Spring 1990, 126–139.

Marshall, George, and Herrick Chapman. *The Social Construction of Democracy, 1870–1990.* New York: New York University Press, 1995.

Marshall, Thomas H., and Tom Bottomore. *Citizenship and Social Class.* London: Pluto Press, 1992.

Marston, Sallie. "The Social Construction of Scale." *Progress in Human Geography* 24, 2 (2000): 219–242.

Marx, Karl. *The Eighteenth Brumaire of Louis Bonaparte.* New York: International, 1987.

Marx, Karl, and Frederick Engels. *The German Ideology.* New York: International, 1972.

McBride, Stephen, and John Shields. *Dismantling a Nation: Canada and the New World Order.* Halifax: Fernwood, 1993.

McClintock, Anne. *Imperial Leather: Race, Gender, and Sexuality in the Colonial Conquest.* New York: Routledge, 1995.

McDonald, Robert. "Working Class Vancouver, 1886–1914: Urbanism and Class in British Columbia." In *Vancouver Past: Essays in Social History,* ed. Robert McDonald and Jean Barman, pp. 33–69. Vancouver: University of British Columbia Press, 1986.

Mehta, Uday. *Liberalism and Empire: A Study in Nineteenth Century British Liberal Thought.* Chicago: University of Chicago Press, 1999.

Merchant, Carolyn. *The Death of Nature: Women, Ecology, and the Scientific Revolution.* New York: Harper and Row, 1989.

Mitchell, Katharyne. "Different Diasporas and the Hype of Hybridity." *Society and Space* 15, 5 (1997): 533–553.

———. "Educating the National Citizen in Neoliberal Times: From the Multicultural Self to the Strategic Cosmopolitan." *Transactions of the Institute of British Geographers* 28, 4 (2003):387–403.

———. "Education for Democratic Citizenship: Transnationalism, Multiculturalism, and the Limits of Liberalism." *Harvard Educational Review* 71, 1 (2001): 51–78.

———. "Flexible Circulation in the Pacific Rim: Capitalisms in Cultural Context." *Economic Geography* 71, 4 (1995): 364–382.

———. "Multiculturalism, or the United Colors of Capitalism?" *Antipode* 25, 4 (1993): 263–294.

———. "Reworking Democracy: Contemporary Immigration and Community Politics in Vancouver's Chinatown." *Political Geography* 17, 6 (1998): 729–750.

Mitchell, Katharyne, and Kris Olds. "Chinese Business Networks and the Globalization of Property Markets in the Pacific Rim." In *Globalization of Chinese Business Firms,* ed. Henry Yeung and Kris Olds, pp. 195–219. London: Macmillan, 2000.

Mitchell, Katharyne, Sallie Marston, and Cindi Katz. "Life's Work." In *Life's Work: Geographies of Social Reproduction,* ed. Cindi Katz, Sallie Marston, and Katharyne Mitchell, pp. 1–26. Oxford: Blackwell, 2004.

Mitchell, Timothy. *Colonising Egypt.* Berkeley: University of California Press, 1991.

Mouffe, Chantal. *The Return of the Political.* London: Verso, 1993.

Mulhall, Steve, and Adam Swift. *Liberals and Communitarians.* Oxford: Blackwell, 1996.

*Multiculturalism and Policing in British Columbia.* Report of the Proceed-ings of the Conference Plenary Sessions Held January 5–6, 1988, Airport Inn Resort, Richmond, B.C. (Available at the Justice Institute of British Columbia Library, New Westminster, B.C.)

*Multiculturalism and the Government of Canada, 1978 Report.* Ottawa: Minister of Supply and Services Canada, 1978.

*Multiculturalism . . . Being Canadian.* Ottawa: Minister of Supply and Ser-vices Canada, 1987.

Münch, Richard. *Nation and Citizenship in the Global Age: From National to Transnational Ties and Identities.* New York: Palgrave, 2001.

Myers, Gustavus. *History of Canadian Wealth.* Toronto: James, Lewis and Samuel, 1972.

Nash, Alan. *The Economic Impact of the Entrepreneur Investment Pro-gram.* Ottawa: Institute for Research on Public Policy, 1987.

———. "The Emigration of Business People and Professionals from Hong Kong." *Canada and Hong Kong Update,* Winter 1992.

Naughton, Barry. "Between China and the World: Hong Kong's Economy before and after 1997." In *Cosmopolitan Capitalists: Hong Kong and the Chinese Diaspora at the End of the Twentieth Century,* ed. Gary Hamil-ton, pp. 80–99. Seattle: University of Washington Press, 1999.

"The Neighbourhood Green." *Cope's 1990 Vancouver Civic Election Report* [election flyer].

Nevitte, Neil. *Value Change and Governance in Canada.* Toronto: Univer-sity of Toronto Press, 2001.

*New Faces in the Crowd: Economic and Social Impacts of Immigration.* Ottawa: Minister of Supply and Services Canada, 1991.

Newman, Peter. *The Canadian Establishment.* Toronto: McClelland and Steward, 1983.

Nonini, Donald, and Aihwa Ong. "Chinese Transnationalism as an Alter-native Modernity." In *Ungrounded Empires: The Cultural Politics of Modern Chinese Transnationalism,* ed. Aihwa Ong and Donald Nonini, pp. 3–33. New York: Routledge, 1997.

Nozick, Robert. *Anarchy, State, and Utopia.* New York: Basic Books, 1974.

Okin, Susan. *Justice, Gender, and the Family.* New York: Basic Books, 1989.

Olds, Kris. "Globalization and the Production of New Urban Spaces: Pacific Rim Mega-Projects in the Late 20th Century." *Environment and Plan-ning A* 27, 11 (1995): 1713–1744.

———. *Globalization and Urban Change: Capital, Culture, and Pacific Rim Mega-Projects.* Oxford: Oxford University Press, 2001.

Olds, Kris, Peter Dicken, Phillip Kelly, Lily Kong, and Henry Yeung. *Glob-alisation and the Asia-Pacific.* New York: Routledge, 1999.

Olsen, Dennis. *The State Elite.* Toronto: McClelland and Stewart, 1980.

O'Malley, Pat. "Indigenous Governance." *Economy and Society* 26 (1996): 310–326.

Ong, Aihwa. "Cultural Citizenship as Subject-Making." *Current Anthropology* 37, 5 (1996): 737–762.

———. *Flexible Citizenship: The Cultural Logics of Transnationality.* Durham: Duke University Press, 1999.

———. "Making the Biopolitical Subject: Cambodian Immigrants, Refugee Medicine, and Cultural Citizenship in California." *Social Science Medicine* 40, 9 (1995): 1243–1257.

———. "On the Edge of Empires: Flexible Citizenship among Chinese in Diaspora." *Positions* 3 (1993): 745–778.

Ong, Aihwa, and Don Nonini. *Ungrounded Empires: The Cultural Politics of Modern Chinese Transnationalism.* New York: Routledge, 1997.

Parekh, Bhikhu. *Rethinking Multiculturalism: Cultural Diversity and Political Theory.* Cambridge: Harvard University Press, 2000.

Pateman, Carole. *The Disorder of Women: Democracy, Feminism, and Political Theory.* Cambridge: Polity Press, 1989.

———. *The Sexual Contract.* Stanford: Stanford University Press, 1988.

Peck, Jamie, and Adam Tickell. "Neoliberalizing Space." *Antipode* 34, 3 (2002): 380–404.

Pettit, Barbara. "Zoning, the Market, and the Single Family Landscape: Neighborhood Change in Vancouver, Canada." Ph.D. diss., University of British Columbia, 1992.

Polanyi, Karl. *The Great Transformation.* New York: Octagon Books, 1975.

Porter, James. "On Multiculturalism as a Limit of Canadian Life." In *The Canadian Alternative: Cultural Pluralism and the Canadian Unity,* ed. Hédi Bouraoui. Downsview, Ont.: ECW Press, 1980.

———. *The Vertical Mosaic.* Toronto: University of Toronto Press, 1965.

Portes, Allejandro, Luis Guarnizo, and Patricia Landolt. "Introduction: Pitfalls and Promise of an Emergent Research Field." *Ethnic and Racial Studies* 22, 2 (2001): 437–453.

Pred, Allan. "Spectacular Articulations of Modernity: The Stockhold Exhibition of 1897." *Geografiska Annaler* 73B (1991): 45–84.

Rafael, Vincent. "Colonial Domesticity: White Women and United States Rule in the Philippines." *American Literature* 67, 4 (1995): 639–666.

Rawls, John. *A Theory of Justice.* Cambridge: Harvard University Press, 1971.

Rekart, Josephine. *Public Funds, Private Provision: The Role of the Voluntary Sector.* Vancouver: University of British Columbia Press, 1993.

Roediger, David. *The Wages of Whiteness: Race and the Making of the American Working Class.* London: Verso, 1991.

Rose, Gillian. *Feminism and Geography: The Limits of Geographical Knowledge.* Minneapolis: University of Minnesota Press, 1993.

Rose, Nikolas. "Governing 'Advanced' Liberal Democracies." In *Foucault and Political Reason: Liberalism, Neo-liberalism, and Rationalities of Government,* ed. Andrew Barry, Thomas Osborne, and Nikolas Rose. Chicago: University of Chicago Press, 1996.

———. *Governing the Soul: The Shaping of the Private Self.* London: Free Association Book, 1999.

Rossbach, Sarah. *Interior Design with Feng-Shui*. New York: Dutton, 1987.

Rouse, Roger. "Making Sense of Settlement: Class Transformation, Cultural Struggle, and Transnationalism among Mexican Migrants in the United States." In *Towards a Transnational Perspective on Migration: Race, Class, Ethnicity, and Nationalism Reconsidered*, ed. Nina Glick Schiller, Linda Basch, and Cristina Szanton Blanc, pp. 22–55. New York: New York Academy of Sciences, 1992.

———. "Questions of Identity: Reflections on the Cultural Politics of Personhood and Collectivity in Transnational Migration to the United States." *Critique of Anthropology* 15, 4 (1995): 351–380.

Roy, Patricia. *Vancouver: An Illustrated History*. Toronto: James Lorimer, 1980.

Rubinoff, Lionel. "Multiculturalism and the Metaphysics of Pluralism." *Journal of Canadian Studies* 17, 1 (1982): 122–130.

Ryan, Mary. "Gender and Public Access: Women's Politics in Nineteenth-Century America." In *Habermas and the Public Sphere*, ed. Craig Calhoun, pp. 259–288. Cambridge: MIT Press, 1992.

Rybczynski, Witold. *Home: A Short History of an Idea*. New York: Viking Penguin, 1986.

Schaffer, Kay. *Women and the Bush: Forces of Desire in the Australian Cultural Tradition*. Cambridge: Cambridge University Press, 1990.

Silverman, Kaja. *The Subject of Semiotics*. Oxford: Oxford University Press, 1983.

Singh, Nikhil. "Culture/Wars: Recoding Empire in an Age of Democracy." *American Studies Quarterly* 50, 3 (1998): 471–522.

Skeldon, Ronald. *Reluctant Exiles? Migration from Hong Kong and the New Overseas Chinese*. New York: Sharpe, 1994.

Sklair, Leslie. *The Transnational Capitalist Class*. Oxford: Blackwell, 2001.

Smart, Alan. "Business Immigration to Canada: Deception and Exploitation." In *Reluctant Exiles? Migration from Hong Kong and the New Overseas Chinese*, ed. Ronald Skeldon, pp. 98–119. London: Sharpe, 1994.

Smart, Alan, and Josephine Smart. "Transnational Social Networks and Negotiated Identities in Interactions between Hong Kong and China." In *Transnationalism from Below*, ed. Michael Peter Smith and Luis Guarnizo, pp. 103–129. New Brunswick/London: Transaction, 1999.

Smith, Neil. "Gentrification, the Frontier, and the Restructuring of Urban Space." In *Gentrification of the City*, ed. Neil Smith and Peter Williams, pp. 15–34. Boston: Unwin Hyman, 1986.

———. "Geography, Difference, and the Politics of Scale." In *Postmodernism and the Social Sciences*, ed. Joe Doherty, Elspeth Graham, and Mo Mallek, pp. 57–79. London: Macmillan, 1992.

———. "Giuliani Time: The Revanchist 1990s." *Social Text* 57 (1998): 1–20.

———. "New Globalism, New Urbanism: Gentrification as Global Urban Strategy." *Antipode* 34, 3 (2002): 427–450.

———. *The New Urban Frontier: Gentrification and the Revanchist City*. London: Routledge, 1996.

————. "What Happened to Class?" *Environment and Planning A* 32 (2000): 1011–1032.

Sparke, Matthew. *Introduction to Globalization.* Oxford: Blackwell, forthcoming.

————. *Hyphen-Nation-States: Geographies of Displacement and Disjuncture.* Minneapolis: University of Minnesota Press, forthcoming.

————. "Networking Globalization: A Tapestry of Introductions." *Global Networks* 1, 2 (2001): 171–179.

Spence, Craig. "Being Neighbourly—By Law." *Vancouver Courier,* December 18, 1985.

Spivak, Gayatri Chakravorty. "Can the Subaltern Speak?" In *Marxism and the Interpretation of Culture,* ed. Cary Nelson and Lawrence Grossberg, pp. 271–313. Chicago: University of Illinois Press, 1988.

Stam, Robert. "Steel of Empire." In *The CPR West: The Iron Road and the Making of a Nation,* ed. Hugh Dempsey, pp. 33–52. Vancouver: Douglas and McIntyre, 1984.

Statistics Canada. *Employment and Immigration Canada, 1985.* Ottawa: Minister of Supply and Services Canada, 1985.

————. *Immigrants in Canada: Selected Highlights.* Ottawa: Minister of Supply and Services Canada, 1990.

Swyngedouw, Eric. "Excluding the Other: The Production of Scale and Scaled Politics." In *Geographies of Economies,* ed. Roger Lee and Jane Wills, pp. 167–176. London: Arnold, 1997.

Taylor, Charles. "The Politics of Recognition." In *Multiculturalism,* ed. Amy Gutmann, pp. 25–73. Princeton: Princeton University Press, 1994.

————. *Reconciling the Solitudes: Essays on Canadian Federalism and Nationalism.* Montreal: McGill-Queen's University Press, 1993.

————. *Sources of the Self: The Making of the Modern Identity.* Cambridge: Harvard University Press, 2001.

Theweleit, Klaus. *Male Fantasies: Women, Floods, Bodies, History.* Vol. 1. Translated by Stephan Conway, Erica Carter, and Chris Turner. Minneapolis: University of Minnesota Press, 1987.

Thompson, John. *Ideology and Modern Culture.* Stanford: Stanford University Press, 1990.

Tickell, Adam. "Global Rhetorics, National Politics: Pursuing Bank Mergers in Canada." *Antipode* 32, 2 (2000): 152–175.

Turner, Roger. *Capability Brown and the Eighteenth Century English Landscape.* New York: Rizzoli, 1985.

Vertovec, Steve. "Conceiving and Researching Transnationalism." *Racial and Ethnic Studies* 22, 2 (1999): 447–462.

Vidler, Anthony. *The Architectural Uncanny: Essays in the Modern Unhomely.* Cambridge: MIT Press, 1992.

Walker, Richard, and Douglas Greenberg. "Post-Industrialism and Political Reform in the City: A Critique." *Antipode* 14, 1 (1982): 17–43.

Walzer, Michael. *Spheres of Justice: A Defense of Pluralism and Equality.* New York: Basic Books, 1983.

Ward, Peter. *White Canada Forever: Popular Attitudes and Public Policy toward Orientals in British Columbia.* Montreal: McGill-Queen's University Press, 1978.

Wardhaugh, Ron. *Language and Nationhood: The Canadian Experience.* Vancouver: New Star Books, 1983.

Ware, Vron. *Beyond the Pale: White Women, Racism, and History.* London: Verso, 1992.

Weaver, John. "The Property Industry and Land Use Controls: The Vancouver Experiences." In *British Columbia: Historical Readings,* ed. W. Peter Ward and Robert McDonald, pp. 426–448. Vancouver: Douglas and McIntyre, 1981.

Webber, Jeremy. *Reimagining Canada: Language, Culture, Community, and the Canadian Constitution.* Montreal: McGill-Queen's University Press, 1994.

Weiner, Martin. *English Culture and the Decline of the Industrial Spirit, 1850–1980.* Cambridge: Cambridge University Press, 1981.

Willems-Braun, Bruce. "Buried Epistemologies: The Politics of Nature in Post Colonial British Columbia." *Annals of the Association of American Geographers* 87 (1997): 3–31.

Williams, Raymond. *The Country and the City.* New York: Oxford University Press, 1973.

———. *Marxism and Literature.* Oxford: Oxford University Press, 1977.

———. *Towards 2000.* Harmondsworth, England: Pelican, 1983.

Wimmer, Andreas, and Nina Glick Schiller. "Methodological Nationalism and Beyond: Nation-State Building, Migration, and the Social Sciences." Paper presented at Dialogue Across the Disciplines symposium, Eugene, Oregon, April 3, 2003.

"Window on the Pacific." *Maclean's,* August 24, 1992, p. 24.

Wiseman, Les. "On Guard for Thee." *Vancouver Magazine,* December 1988.

Woodsworth, James. *Strangers within Our Gates.* Toronto: University of Toronto Press, 1972.

*Working Together towards Equality: An Overview of Race Relations Initiatives.* Ottawa: Minister of Supply and Services, 1990.

Yeoh, Brenda, and Kate Willis. "'Heart' and 'Wing,' Nation and Diaspora: Gendered Discourses in Singapore's Regionalisation Process." *Gender, Place, and Culture* 6, 4 (1999): 355–372.

Yeoh, Brenda, Shirlena Huang, and Katie Willis. "Global Cities, Transnational Flows, and Gender Dimensions: The View from Singapore." *Tijdschrift voor Economische en Sociale Geografie* 91, 2 (2000): 147–158.

Yeung, Yue-Man. "Hong Kong's Business Future: The Impact of Canadian and Australian Business Migration Programs." In *Pacific Asia in the 21st Century: Geographical and Developmental Perspectives,* ed. Yue-Man Yeung, pp. 309–339. Hong Kong: Chinese University Press, 1993.

Zizek, Slavoj. "Multiculturalism, or the Cultural Logic of Late Capitalism." *New Left Review* 225 (1997): 28–51.

Zukin, Sharon. *Loft Living: Culture and Capital in Urban Change.* Baltimore: Johns Hopkins University Press, 1982.

# Index

*Note:* Page references followed by *f* and n refer to figures and notes, respectively.

race: French and British as, 96, 236n25;
    use of term, 221n6
race relations, between British and
    French Canadians, 96
racism, 80–83; in Britain, 195–196; capi-
    talism linked to, 177; cultural con-
    flict conflated with, 164, 194–197;
    definitions of, control of, 84; in
    demonstrations, 80–81; developers'
    allegations of, 82, 83; education cam-
    paign against, 105; Festival '91 and,
    53–54; government organizations
    opposing, 106–111; Humphreys
    (Annie) on, 176–178; immigrant reac-
    tions to, 193–199; among immigrants,
    205; in Kerrisdale, 80–81; media cov-
    erage and, 71, 72–73, 82; multicul-
    tural efforts to address, 102–111;
    multiculturalism as form of, 93; in
    neighborhood organizations, 81–83,
    84–85; perception of speculation and,
    68; property prices and, 108; rise in
    Vancouver of, 71–72; in Shaughnessy
    Heights rezoning efforts (1992), 140,
    148–154; of slow-growth movements,
    178; in social liberalism, 5–6, 84; as
    social vs. financial issue, 177; spatial
    segregation based on, 167–168; strate-
    gic use of, 10, 82, 83; in tree removal
    conflict, 158–159; wealth emphasized
    in, 71; in zoning regulations, 141–
    147
radio, 72
Rafael, Vicente, 165
Rankin, Harry, 130
rationality: of feng shui, 200; and politi-
    cal participation, 216; universal, 25
rationalization, of land governance
    issues, by media, 134–135
Rawls, John, 29, 214
real estate. See property
real-estate companies, 69. See also
    developers
Real Estate Weekly, 108
reason: exclusion based on, 25, 26–27,
    30; familiarity linked with, 30
recession, economic, of 1980, 45–46
redistribution: neoliberalism undermin-
    ing, 33; push for, 28–29. See also
    welfarism
Reform Party, anti-immigration plat-
    form of, 166

rent, increases in, 42–43
reproduction. See social reproduction
Repton, Humphry, 186
Residents Save Vancouver Please
    (RSVP), 82
respatialization, vs. deterritorializa-
    tion, 8
reterritorialization, deterritorialization
    linked with, 8
revanchist movement, 40, 41, 42
revisionist liberalism, 28–30; classical
    liberalism critiqued by, 30; national-
    ism in, 28; and state legitimacy, 29;
    tenets of, 29
Richmond, Claude, 56–57
Richmond (suburb): development activ-
    ity in, 191; middle-class immigrants
    in, 204–205, 248–249n69
rights. See individual rights; property
    rights
Roberts, Andrew, 140
rootlessness, 156
Rose, Nikolas, 33
Rouse, Roger, 89, 90
Royal Bank of Canada: in Laurier Insti-
    tute, 110; services provided by, 64
Royal Commission on Bilingualism and
    Biculturalism (1963), 98–99
Royal LePage, 65, 69
Royal Trust Asia, Ltd., 65
RSVP. See Residents Save Vancouver
    Please
Rubinoff, Lionel, 98

Samuels, Jane, 188–189
scale: dominant, of governance, 9;
    social construction of, 9–10
Seattle Times, on regional development,
    133–134
Seelig, Michael, 132–140
self-orientalizing, 10
Semi-Tech (Global), Ltd., 53
senior citizens. See elderly women
separatism, French, 98
sequoias, removal of, 155, 156, 158
The Sexual Contract (Pateman),
    25–26
Shaughnessy Heights, 76, 144–154;
    1982 rezoning of, 146–147; 1992
    attempted rezoning of, public hear-
    ings on, 140, 148–154; Canadian
    Pacific Railroad land in, 141, 144;

*tai hong yan*, 59
taxation: head, on Chinese immigrants, 246n18; of property development, 110
Taylor, Carole, 154
Taylor, Charles, 118, 120–121, 214, 238n77
Taylor, Peter, 72
TEAM (The Electors Action Movement), 40
technological innovations, and transmigrant laborers, 15
territory: in classical liberalism, 85; in cultural identity, 90–91; and hegemonic formation, 18; in multiculturalism, 88, 91–92; national, vs. state policies, 18; in political identity, 85, 214–215; vs. space, 8. *See also* deterritorialization
Thatcher, Margaret: racism under, 196; in Sino-British Joint Declaration (1984), 4
Thatcherism, hegemonic formation underpinning, 18–19
Theodore, Neil, 224n58
*There Ain't No Black in the Union Jack* (Gilroy), 166
Theweleit, Klaus, 208
Tickell, Adam, 33
Tienanmen Square massacre (1989), 4
TNCs. *See* transnational corporations
Toigo, Peter, 48
Toronto Caravan, 97, 100
Town Planning Act (1925), 145–146
Townsend, John, 245n10
tradition, 185–192; British, 185–190; in homes, 185–190; letters to City Hall on, 187–188; letters to editor on, 191–192; old vs. new money and, 190–191
tramway service, in Point Grey, 145
"TRANS/PLANTS: New Canadian Entrepreneurs" (business fair), 159–160
transmigrant laborers, 14–15; "home" of, 14–15; nationalism among, 15
*The Transnational Capitalist Class* (Sklair), 221n12
transnational corporations (TNCs), rise of, 13
transnationalism: Business Immigration Program contributing to, 59; as choice, 14, 15–16; families in multiple locales with, 210–211; in global-

ization, 11; liberalism challenged by, 6, 216; of managerial labor, 14; multiculturalism and, 89, 91–93, 119–120, 218; space in, 8; unfamiliarity in, 31–32
transplantation, imagery of, 160
transportation, modernization of, 46–47, 48
tree removal, 154–161; in feng shui, 201; government regulation of, 158–159; imagery of, 156–157, 158, 160–161; media coverage of, 158; as moral issue, 156–157, 244n74; organizations opposing, 155, 244n81; racism and, 158–159
Trudeau, Pierre, 40, 97, 99–100
trust companies, in property investments, 64–65
trusts, offshore: legalization of, 231n73; in United States, 63
Tudor style, 186, 187f

uncanniness, 211, 249n81
unemployment rates, in recession of 1980, 45
United Chinese Community Enrichment Services Society. *See* SUCCESS
United States: Canadian identity as distinct from, 94; philosophical differences with Canada, 87, 235n3
universal rationality, 25
universalism, failure of, 219
universalization, of land governance issues, by media, 134, 135–136
uprootedness, 157, 160
urban planning. *See* land governance
Urban Planning, Vancouver Department of, in tree removal conflict, 158–159
urban policy, Vancouver: in 1970s, 40–41; in 1980s, 41–42; control of, 40–41; provincial government role in, 40, 44; shift from social liberalism to neoliberalism in, 40
urban sprawl, threat of, 136

vacancy rates, 42, 227n8
vagabond capitalism, 78
Vancouver: map of greater, xiv; map of local, 77
Vancouver Land and Improvement Company, 145
Vancouver Land Corporation (VLC), 49

KATHARYNE MITCHELL is Professor of Geography and the Simpson Professor of the Public Humanities at the University of Washington.